IN THERE SOME

Including a Biography of
Steven Allen
Master Craftsman,
Dry Stone Waller

Written and Illustrated by
David Griffiths

In There Somewhere

Written and illustrated by David Griffiths© and published by the Dry Stone Walling Association of Great Britain.

First edition published 1999.

The Dry Stone Walling Association of Great Britain is a registered charity (registration number 289678).

All proceeds from the sale of this book are used to further the work of the Association.

**The Dry Stone Walling Association of Great Britain,
Registered Address: c/o YFC Centre, National Agricultural Centre, Stoneleigh Park, Warwickshire CV8 2LG.**

All rights reserved. No part of this publication may be reproduced, stored in a retrieval system, or transmitted in any form or by any means: electronic, mechanical, photocopying, recording or otherwise, without the prior, written permission of DSWA.

Printed by G W Belton Ltd., Gainsborough, England

ISBN: 0 9512306 5 4

Definition: dry stone wall

A wall, built from unmortared stones,
held together by itself.

Dedication:

I dedicate this book to all
those who take pride and
joy in building dry stone walls
and who fight tirelessly for their
preservation.

Acknowledgements:

I would like to thank the DSWA for their help and support in the preparation of this book; to Jacqui Simkins for her invaluable help, skilful editing and unflagging encouragement particularly at the proofing stage; to my own Yorkshire Branch of the DSWA for their unconditional support and friendship; to Andy Goldsworthy for giving me access to his work on the Sheepfold and other sculpture projects in dry stone; and finally to Steven Allen, without whom this particular book would not have been written.

David Griffiths

FOREWORD

Having been brought up in the Cotswolds and lived for many years in Derbyshire, I cannot imagine a country without stone walls, so it is with the greatest of pleasure that I write this foreword.

There is a great interest now in the rural crafts and skills which for too long have been taken for granted. High among these is walling. Lately people from all walks of life have become fascinated by it and are learning how this age-old method of enclosure is done, the slow and painstaking work which defies the weather and stands the test of time for hundreds of years.

David Griffiths explains it all from the "footings" to the "copings". His chapter on Competitive Walling sets down the way that quality shines out in the work of the experts. It reminds me of when I took a friend to see a walling competition near Chesterfield. He had never been to one before. He is a Londoner responsible for perhaps the most famous antique shop in Bond Street, his speciality being 18th century furniture – a far cry, you would think, from dry stone walling. When the competitors had finished their stints, my friend looked at each one with great care. He quietly told me his choices of first, second and third. After the judging was completed, the results were as my friend forecast. That taught me than an eye for perfection covers many skills!

The tale of Steven Allen is an engrossing one. His progress from practical walling for farmers to sculpture in the same medium for the imaginative Andy Goldsworthy is described so well that people will feel obliged to go and see for themselves the results of the labours of these two exceptional men. Their walls from Cumbria sheep folds to the "walking walls" in America cover the spectrum of this way of building, which is an essential ingredient not only for the farmers but for those who find delight in views of the most dramatic country in these islands. Without the walls the impact of the upland scenery would be sadly diminished.

I commend this book to all who admire the skill and beauty combined and who take an interest in the country in which we live. Through it we will look at the familiar with new eyes.

Deborah Devonshire
Chatsworth 1999

FROM THE CHAIRMAN OF
THE DRY STONE WALLING ASSOCIATION

Over a period of two years I observed David Griffiths as he carried out research for this book. I was intrigued to find he had chosen one of the most respected of the Dry Stone Walling Association's Master Craftsmen as the vehicle on which to show the variety of activities that the modern dry stone waller may be involved with.

This book explores the agricultural roots of the craft, through landscaping and garden projects to modern-day artistic use of the craft in construction of major sculpture works.

David is much respected as an author and artist but he is also a competent exponent of the craft of dry stone walling. I have watched David develop from a raw novice into an extremely skilled amateur waller who has honed his skills in competition throughout the United Kingdom. It is this unusual combination of talents that has produced a superbly illustrated text of interest to all.

I am pleased that Her Grace The Duchess of Devonshire, the Patron of the Dry Stone Walling Association has provided the foreword for this book, as she is a great supporter of the craft.

As Chairman of the Dry Stone Walling Association of Great Britain I recommend this book to everyone who is concerned with or about this ancient craft, conservation and our heritage.

Paul Webley

LIST OF ILLUSTRATIONS

Fold at Mungrisdale	1		Coping stones	63
Gap in wall	16		The finish	64
Repairing gap	18		Basic wall structure	66
Wall surrounding copse	23		Footing	67
Entrance to farm	24		First lift	68
Circular fold	26		Throughs	68
Rectangular fold with cone	26		Second lift and throughs	69
Casterton fold	27		Retaining wall	70
Cone	29		Wall on steep slope	71
Fold at St Albans	31		Curved wall	72
St Albans' fold rebuilt	32		Cheekend	73
Mungrisdale fold	33		Corner	74
Boulder in wall, USA	34		Lunky	75
New York project	35		Stile	75
Walking wall, USA	36		Steps	76
Whisky bottle	38		Pillar base	76
Bungalow at Greystoke	40		Arch for lunky	77
Cumbria garden	41		Pillars and arch	78
Garden perimeter wall	42		Cumbria wall	81
Long roadside wall	43		Langwathby wall	82
Roadside wall with stoup	44		Pennine wall	82
Entrance at Oasis	45		Cotswold wall	83
Sculpture at Heathrow	47		Cornish wall	83
Chelsea Flower Show	48		Clawdd	83
Chelsea exhibit planted	49		Galloway dyke	84
Stripping out	58		Irish wall	85
Footings	60		Cumbria flag wall	86
First lift	61		Welsh slate fence	86
Throughs	62			

CONTENTS

Foreword by Her Grace, Duchess of Devonshire   7

Message from Chairman of DSWA   9

Introduction   12

Chapter One: 15
Steven Allen, Master Craftsman, Dry Stone Waller

Chapter Two: 26
Craftsman and Artist
Andy Goldsworthy Projects

Chapter Three: 50
Competitive Walling

Chapter Four: 65
Master Craftsman Certificate

Chapter Five: 79
Walls and walling
A Survey of styles and condition in Britain

Chapter Six: 87
In Conclusion

Appendices:
  1 Glossary of terms   90
  2 Bibliography   93
  3 List of organisations   95
  4 DSWA Statement of policy   97
  5 DSWA Craftsman Certification Scheme   98

INTRODUCTION

When I am looking intently at a wall - usually as I am drawing it - I find myself constructing it. I try not to cheat in an impressionistic way. Because of the way it was built, the wall has to fit together technically on my page and reveal the tampering, repairs and climates of centuries - or not as the case may be.

For most of my working life I have lived in areas of city and village where land and routes are marked in stone. I patched my own first tumbled dry stone wall without any real knowledge of how it had been originally assembled. The urgency to do this was compounded by the need to keep my children enclosed within the garden perimeter and to keep out the inquisitive heifers from the adjacent field.

Since then I have become passionately obsessive about constructing in stone and in particular about the dry stone walling styles which can be found in Britain and in other parts of the world. I have also become increasingly concerned at the way in which such walls are being systematically demolished, stolen and mismanaged.

More recently, I wanted to write and illustrate a comprehensive book about dry stone walling but had been unable to find the catalyst that would shape my experience and focus my attention upon present-day wallers and the conditions in which they practice their art and - more realistically - earn a living.

I first saw Steven Allen walling three years ago in the final event of a Grand Prix[1] competition held at Lawkland Green Farm, Austwick near Settle in North Yorkshire. It was a warm, dry, September Saturday. I also remember the day and the location because my young son could walk and play safely without constant supervision in the open spaces surrounding the competition wall and I was able to concentrate for long periods of time upon the competition - the first I had seen.

---

[1] **Grand Prix:** National walling competition of the DSWA

Steven had drawn a stint[2] next to a mature oak. The disadvantage being that he was instructed to leave a gap between the end of his wall and the tree - an access point for the competitors. This meant that he had to secure one end of his wall roughly in the way a cheek end[3] is constructed. He had nothing to tie into and it would look unfinished. The advantage was a greater choice of stone due to the tumble[4] around the tree being technically part of his stint.

At first I walked up and down each side of the wall making mental notes about how it was being constructed, and the standard of walling in the professional, amateur, novice, and junior classes.

I found myself eventually drawn to Steven's work. There was something compelling about the way he walled. Despite the fact that he always seemed to be ahead in terms of the amount of stone that he had assembled, the work looked effortless. It was though the wall was already constructed in its pile on the earth and all he had to do was lift it into place. Despite his height (he stands 1.8 m) the movement between the pile of stones and the wall was so smooth you didn't see it.

He looked much younger than the rest of his professional competitors. I didn't see him speak. He seemed wrapped in a 'do not disturb' parcel of concentration. Whilst some around him flashily clattered stones with their hammers, he would only occasionally use his to chip off a small protrusion to create a tighter joint.

Although the horizontal lines of construction seemed to flow easily along the wall in reality these 'lines' were broken regularly because of the randomness of the stone. I was seeing a straightness that wasn't there. When I compared his wall with others I noted that his looked 'tidier'; I couldn't express this at the time in any other way.

At the end of the day the judges awarded him first prizes in all the professional categories. Not only was he declared National Champion and outright winner of the Grand Prix competition, but

---

[2] **Stint:** Measured length of wall for each waller in competitions

[3] **Cheek:** A system of construction that seals and ties the ends of a self-supporting wall

[4] **Tumble:** Where part of a wall has fallen or "tumbled"

he won the trophies for being the best professional and for building the best wall on the day. This was the first time in the history of the competition all the trophies had gone to the same waller.

When I finally decided to write this book I hoped that Steven's consummate craftsmanship would be represented at the heart of it. I didn't want to write the story of someone long dead whose life was recorded in anecdote and the myths of romantic memory, I wanted to make real contact with a living master craftsman and to record his story whilst he was at the height of his walling career.

For me, Steven is one of the finest exponents of the craft of dry stone walling of his generation and this personal judgement is acknowledged as such by his fellow professionals.

Fortunately he allowed me to watch and work with him on an extraordinary range of walling projects and it was during the hundreds of hours we spent together that I was able to glean his technical, experiential and sometimes philosophical views about walls and walling.

He has given me unlimited access to his own personal journals and he, his wife Susan and his daughter Hannah have always provided the very best of Cumbrian hospitality.

Writing and illustrating this book has been a privilege and a joy.

CHAPTER ONE

Steven Allen

Master Craftsman, Dry Stone Waller

Introduction

In areas where hedging and fencing prevail, such as the flatlands of Lincolnshire and Cambridgeshire, the early patchwork of boundaries have been long removed and replaced by unbroken prairies created mainly for the production of grain.

Just as the physical landscape has dramatically changed to accommodate this philosophy so the ecological and human occupation has altered accordingly. It would be almost impossible for a young farmer at the end of the twentieth century, living and working in these flatlands to conceive how his local environment must have looked only forty or fifty years ago let alone imagine his daily and seasonal work within it.

Thankfully, such evolution has hardly touched the many rural areas of mountainous Britain where the soil is thinly spread over beds of rock. Most of the land is still marked in a patterned, walled division and clearance completed centuries ago.

Such an area is Cumbria in the North West which embraces some of the most dramatic and picturesque mountainous landscape in England. The Lake District is one of the largest and most popular tourist areas in Britain. It also supports a farming community skilled in the raising of livestock - mainly sheep - who scratch a living using skills handed down from generations of forefathers in a bleak, barren, walled terrain which has remained mainly untouched since it was created.

It is within such a landscape and tradition that Steven was born and has continued to live and work.

Gapping (an aside)

Gap in wall

As walls divide most of the local landscape, maintenance is a necessary feature of farming life. Gapping[5] is 'bread and butter' work for most wallers and many would agree that working on this basic repair is where most of them began their introduction to their craft.

The main technical conditions which prevail at a 'gap' in a wall is that either side of the fall usually retains much of the cross-section of its original construction. Also the stone which has tumbled is readily available to re-use in the repair. It doesn't take too much working out that if the layers of stone, and the angle of each side of the wall is copied then the gap can be notionally rebuilt to its original design.

In practice this is not as easy as it looks. The reasons for the tumble, which created the gap in the first place, have to be resolved so that the problem is not repeated. A wall falls for many reasons and those reasons become obvious as the gapper strips the rocks down to the lowest layer of strength.

---

[5] **Gapping:** Repair of that part of a wall that has fallen

The most usual reasons for tumbles are generally because the footings[6] have been poorly laid and have moved; the layers of stone have been poorly tied;[7] the amount of hearting[8] is inadequate or poorly set; or the top stones or copes[9] have not adequately spanned the wall, been too loosely placed, or have become dislodged. Sometimes the cause is more obviously the fact that a tree has grown close to or within a wall and its development has taken the wall with it.

In a well built wall, if one stone covers the joint of another on the course or layer below it 'ties' or holds that joint in place - and so on. 'One over two and two over one' is a familiar expression used by wallers to describe this obvious technical description of 'bonding'. However, if lines of joints are placed directly over each other they already begin to look like the cracks they will eventually become.

Usually, the ugliest and largest stones are buried in the footings and the size of stone on each course reduced as the wall grows upwards. Wherever possible 'throughs'[10] should span the wall at regular metre lengths apart. If this is not possible - as in many Cotswold walls - then the fillings should be placed with even greater care. Sometimes cement is thrown into the heart of the middle two or three courses of a Cotswold wall to creating an artificial 'through' system.

If lines are pegged across the gap, then it is likely, even if the construction is random[11] in style, that the lines will guide the eye to maintain the line of existing courses and follow their 'batter'.[12]

---

[6] **Footings:** Foundations

[7] **Tied:** The way in which one stone overlaps, or covers, another above and below it to make them securely bonded

[8] **Hearting or filling:** Small stones that are used to pack and support the centre of the wall and is a crucial part of dry stone walling construction

[9] **Copes:** Stone that span and tie together the top of the wall

[10] **Throughs:** Large slabs of stone that span the width of the wall and help to tie it together

[11] **Random:** All the stones used are of an obvious irregular shape and size and the style reflects this in the look of the face of the wall

[12] **Batter:** This refers to the straightness of the wall along its length and the vertical angle of the sides between the footings and the cope

As well as seeing what the final wall should look like and having a cross-section readily available to refer to on either side of the gap, the novice waller soon learns how to organise the stone and handle it during reconstruction.

Stones are picked up, tried, rejected and placed, and it soon becomes clear that quite often the problem of selection can be done from the ground before they are moved. Lifting and handling stone demands effort as well as skill. Sorting stone into rough sizes on the ground before they are lifted into place can save a lot of energy.

Repairing gap

Learning to fill gaps provides a good introduction to acquiring the basics of walling practice. Most wallers complete their 'apprenticeship' repairing gaps. To do this requires matching the often-uneven line of the wall with a strong reconstruction. From an early age Steven had a special aptitude towards making sound repairs which 'looked right'. Most of all he enjoyed doing it.

Sometimes walls have to be repaired urgently because of the need to contain livestock in particular fields or collecting pens, especially sheep, which are notorious for being able almost instinctively to find weaknesses in walls. So learning how to rebuild an existing wall quickly and for it to be technically strong were the first walling skills that Steven learned. **Strength. Style. Speed.** The three elements that form the basis of the craft of the professional dry stone waller. The fixing of these skills would only become instinctive with practice, and Steven had plenty of that!

Even in his present secure position as a much sought-after waller, he rarely turns down this very basic repair work. He is only too aware that years of regular 'gapping' on his father's farm provided him with a thorough apprenticeship for his future walling career and he determines never to be dismissive of this basic walling practice.

Childhood to professional waller

From the age of about two until the age of eight, Steven lived on a small farm at Greenholme near Tebay in the heart of Cumbria, which – coincidentally - is about three miles from where he lives now. A garden wall surrounded the house and Steven's earliest recollections were of his re-arranging those topstones[13] on this wall he could manage to move.

He can't remember ever being 'taught' by his father Gordon but he remembers watching him repair walls from this early age and sometimes helped him when he came home from school and during holidays.

His father would produce good, strong repairs, but he had no extra interest in the craft of walling apart from what was a necessary part of farm maintenance. This, according to Steven, is the environment in which most wallers develop their first contact with walling, learning what is necessary to satisfy a general knowledge of boundary or building repair. The majority of professional wallers recall this as the time when they develop the basics of their craft. 'Farmers, farmers' sons, farm workers ... they were brought up with it.'

Steven remembers a wall that used to start close to the back lane beside the farm and stagger in a tumbled state up the next valley. When he appeared to have nothing to do his father used to encourage him to go and 'play' on it. The wall has practically disappeared now but Steven remembers it, and especially that the stone was small and easy to handle. He was seven at the time.

The family then moved about three miles to a bigger farm at Brockholes. Steven has no recollection of his walling skill developing significantly over the next few years - of his early

---

[13] **Topstone:** Cope

primary schooling. But he recalls clearly the first times at around the age of thirteen or fourteen, working alongside his father collecting and helping with the filling between the stones his father had placed.

He would also go round the walls before silage[14] time picking up and replacing stones that had fallen from field walls. This was not gapping and didn't happen every day, but occupied him after school. In the village there was 'not much to do'.

He left school when he was nearly sixteen and went to work immediately at a farm at Shap, his father's farm being hardly big enough to realistically sustain the need for two full- time workers.

One of his early, familiar tasks was to pick up stones and replace them on field and boundary walls, including gaps, the difference being this time that he completed the repairs without assistance or supervision. He was told to 'go along there and see what he could do' with the roughish limestone - and he did.

Satisfied that he could do a reasonable job his employer encouraged him to tackle some of the major wall repair work on his property. He did this occasionally in the general scheme of farm work over a period of about eighteen months. Steven recalls that throughout this time he didn't use a building line or a hammer - and he didn't possess either. His was a ' thrown in at the deep end, and learn as you did it', apprenticeship.

In 1978, in his eighteenth year, he returned home, and became immediately involved with the construction of new buildings about the farm. He also began to repair gaps having convinced his father that he could do so. Bigger and bigger repairs challenged his development as a waller but he still worked without a hammer or a line.

When the building programme was finished he started working on the neighbouring farm part-time, two days a week. It coincidentally happened to be a farm having miles and miles of walls and he was quickly employed on the familiar repair routine.

---

[14] **Silage:** Fodder crop of grass, harvested while green, vacuum stored in state of partial fermentation.

Towards the end of the next ten years most of his working time was spent repairing or building dry stone walls.

From the beginning of this period Steven developed a real interest and - more importantly - a continuous enjoyment of walling. Whether this was reflected in his attitude or not, by the time he was twenty he was spending most of his working days on his neighbour's farm gapping.

It was also significant that he joined the Young Farmers Club[15] during this time. Significant because he became exposed to their social activities and in particular their competitions. In 1984, on the 25th April, they held a Cumbrian Young Farmer's walling competition at Flookborough near Grange-Over Sands.

He used a hammer for the first time in the competition. His father found one from somewhere and gave it to him with the immortal words: 'You're going in for this competition. You may need this.'

Throughout the day he was conscious of assembling stones to make a solid construction, trying at the same time to make them 'look good'. He 'just did it'. It seemed common sense to put the biggest stones at the bottom and grade the rest of the stones to the smallest at the top, placing the throughs where he thought they ought to go. As simple and as ingrained as that.

At the end of the competition he felt his wall to be as good as the others. The judges went one step further and awarded him first prize. He never expected to win. Even now, when competing, he still does not anticipate victory.

Despite his success Steven was not moved to alter his workstyle or place. It was 1985 and he was about to get married to Susan and he felt this was not the occasion to be changing the financial security of working part time on three different farms.

However, this was the time when he developed a real interest in competitive walling. His competition scrapbook begins in 1986 and

---

[15] **Young Farmers' Club:** YFC organisations are self-governing bodies that promote activities for the youth who work within or who are part of the rural community in Britain

records his certificate of completion at Moorcock Show at Mossdale (at which HRH the Prince of Wales was present) and The Friends of the Lake District Competition a year later.

His first competition in the professional category was at the Dufton Show in 1987 where he was awarded fifth prize. He was slightly aggrieved at this positioning and felt that he should have been awarded third prize. He felt that he'd been categorised as a novice professional and placed accordingly. Steven had quickly to learn to accept the subjective opinions of judges and that sometimes it would go in his favour at the expense of others with similar grievances. This is the way of all judged competition. But it revealed a keen competitive edge. A very determined will to win.

Over the next two years he was placed in the first three in all the competitions he entered including a first at the Penrith Show before winning the prestigious Friends of The Lake District Competition in 1989. The significance of this victory was that the competition received media coverage and Steven found himself the centre of reportage in the local press.

This was the moment when he began to think about walling for a living. Until this point in time he had never walled for a fee outside his normal farm work. However, whilst competitive walling fulfilled a recreational pleasure it also introduced him to the walling skills of others who were working professionally full time.

Instinctively he measured his walling skills against those he'd seen working locally and in competitions and was amply satisfied that his standards were up to those of the professionals. The biggest problem was there were no local, full-time professionals to whom he could refer for advice concerning the pitfalls of starting up his own dry stone walling business.

He went further afield competing and winning at places like Staffordshire and Chesterfield before returning closer to home to win at Dufton, Moorcock and gain a second at Penrith.

Inevitably he had the opportunity to talk with professional wallers and discovered the not too secret fact that some were earning considerably more per week from walling than he was earning working on three farms, and - the biggest carrot of all - they ran their own business.

Despite the fact that Susan was now pregnant and would soon have to take maternity leave, he began to think very seriously about making the big step of becoming self-employed.

In 1987, Hannah was born. This wonderfully significant moment moved Steven, now 27, to make his first tentative step to begin his own, one-man dry stone walling business. Susan, as ever, totally supportive, offered her skills to help with the administrative side of the business.

Wall surrounding copse

It was in 1988 that Steven decided to advertise his walling skills in the local press and, if his request for work was successful, he would give up one of his farming jobs for two days a week.

Within two days he had secured a contract to rebuild 300 metres of tumbled wall surrounding a copse for a local farmer. He was to be paid in two instalments: the first, at the halfway point of construction, and the second at its completion. He began in March and received his first payment in June.

Entrance to Farm

He was immediately contracted to build walls either side of the entrance to the same farmer's domestic property. This became a more public project and received a lot of local attention during its construction and after its completion. He recalls that he got a lot of work from that wall. In fact the same farmer and most of his neighbouring farmers are still his customers today. It was during this same year (1988) that Steven decided, very quickly, to take up walling full time.

Even in these early days as a professional he was never inclined to 'throw a wall up'. He always strives to maintain the highest standards of strength and style. That is his trademark. Anything less will produce a rapid reduction in contracts. This philosophy is taken on when working out the cost to customers - and yet this costing still has to be competitive.

He almost always charges by the metre not the time taken to construct. Charging by the metre fixes the estimate and puts the responsibility for its completion to the correct standard at the agreed cost within a given time solely upon the waller. It is an exacting and demanding contractual arrangement. Time can be fiddled as many who have employed 'cowboy' wallers can testify.

From this solid beginning, Steven has never been without continuous work and he has sustained a good living from walling. He has been blessed with continuing good health and has been injury free. He has also been courageous in that he has taken on many new and different projects that others, less confident in their ability, would have refused.

Much of this more recent work has been commissioned as a result of his many competition successes. Winning in every corner of walling Britain, becoming the National Professional Champion no less than four times, has been of paramount importance in enhancing his reputation. Work has followed in hot pursuit of his competitive walling successes.

It is to some of these exacting, exciting and most extraordinary projects that I now want to turn my attention, for they represent some of the finest examples of the craft of dry stone walling, and they stand as a flagship, advertising this fact. They have helped to create a recent upsurge in interest and enquiry about the continuous development of dry stone walling and its use in works of art.

CHAPTER TWO
Craftsman and Artist
Andy Goldsworthy Projects

Circular fold

It is ironic that when I began the research for this book Steven had just become involved in the Andy Goldsworthy Sheepfold Project.

The five year, Millennium project, funded by Cumbria County Council, Northern Arts and district councils, is centred on the construction of 100 sheepfolds, washfolds and pinfolds located on the fells, beside drove paths, in villages and settlements in various parts of Cumbria. Where the originals were designed to contain and protect sheep, the millennium versions will, in some cases, additionally embrace either huge fieldstones gathered from the surrounding landscape or specially designed, dry stone cones.

Rectangular fold with cone

They will have entrances that can be walked, stepped or peered into, host the surprises already described, or simply be empty, contained spaces of stillness and protection.

The fact that many of the folds will be reconstructed on the sites of the originals is part of Goldworthy's play with time, connecting the new walls with the old and today's craftsmen with their predecessors. Just as the walls are built in 'layers' of the indigenous stone, so the landscape and its history is physically and socially 'layered'.

Goldsworthy refers often to these 'layers' linking the conceptions of physical strength, the effect of light upon the angularity of surface and how it can be changed with a fresh vision of its re-assembly. It is how he uses the folds and shifts the emphasis in his additions without disturbing their general outline, which reflects his own sensitivity and autobiographical links with the natural environment in general, and hill farming in particular.

Some of the folds, already completed, growing from the hillside earth, stationed in corners of pasture or open moorland behind walled drovers' pathways are, in my view, simply breathtakingly beautiful.

The first journeys I made from Leeds to Cumbria to meet with Steven led me to the early sheepfolds being constructed beside the old drovers' path along the Fellfoot Road close to the village of Casterton, two miles north east of Kirkby Lonsdale.

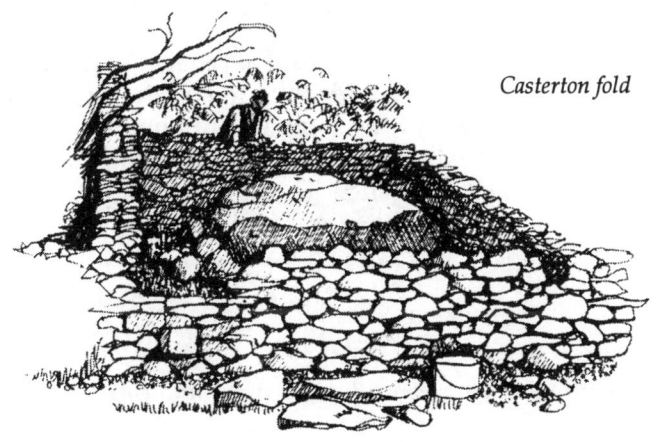

Casterton fold

The folds under construction are sited in the corners of existing walls which divide the fields on one side, and the bracken and fern - wooded slopes climbing to the hill tops on the other. Access into the folds from the path is by steps leading through a traditional 'U' or 'V' shaped 'step-through' stile.[16] The steps are, almost always, old stone gate posts which have, through time, frosted and broken at the point they have been set in the ground but still retaining enough length to span the wall like a 'through' and produce a wide, solid step on either side.

Not only do these entrances provide access and egress to the fold but they make a head- height window offering a view of inclusions or otherwise within. Sometimes, when the fold is built at the same horizontal level as the adjoining field, the dark outline above the cope, turns out to be a huge boulder, set in the centre of the interior. Looking at it from the stile, stepping into and walking around it affects your perspective of it and, in consequence, its ever-changing presence.

Other folds, which tilt with the hillside towards the path, openly reveal their inner monoliths[17] without the need to peer through their stiles.

Each fold is rectangular and is built in the style of the existing walls that border the path and divide the landscape surrounding it. Sometimes, if they are in good enough condition, the existing walls - which make up one or two of the four sides of the fold - are left undisturbed and patched to make up the correct height, while others are either pulled down to their footings or rebuilt from new.

Their presence remains unobtrusive and serenely quiet. You have to look for them to discover them despite the obvious repair and straightening of the walls and the steps leading into them. They are not sculptures that ignore the landscape in which they are set or shout with a 'look at me' arrogance. They are part of its evolutionary development and make a real sensitive contact with the social and ecological history surrounding them.

---

[16] **Stile:** A structure that enables walkers to step comfortably through or over walls

[17] **Monolith:** monument or sculpture consisting of a single standing stone

This does not mean that their presence has not produced controversy amongst some local farmers and passing visitors. Steven is often confronted with comments from walkers as he is working on the folds. He finds it difficult to defend the expenditure on sculpture when this funding could have gone on education, and the health service.

However, he always expresses his personal delight and privilege at being so closely involved with the project as it provides him with a demanding challenge to extend the boundaries of his and other wallers' skills and experience. He also points to the fact that the folds have attracted much interest from tourists, and that this, plus the purchase of materials from farmers and the extra work which has been brought into the area has benefited the local economy. He likes the work as art and says so, publicly. He is passionate about his craft and sees this work as an exciting development of it.

Because the folds are built to such a high specification they represent the best of the master craftsman's skill. Because Andy Goldsworthy has such an international profile and respect as a sculptor then the work in dry stone shares the attention he receives. Rebuilding, renovating and creating new dry stone walls protect tradition, landscape and work for future generations of wallers. Such a public demonstration of the very best of the craft - albeit as a work of art - provides one of its best possible means of public display. It can only be good for wallers and walling.

Cone

The process of building the folds follows a regular pattern. Once the site has been chosen, agreements negotiated with the local landowner, stone gateposts (stoops) and extra stone collected, Andy Goldsworthy pegs out the foundation perimeter of the pens - often around a huge stone. This stone will have been selected from the many that rest on the surrounding hillside and moved with use of a JCB to sit in the fold site.

If the fold is located in the corner of a walled boundary, two of the existing walls will suggest a rough blueprint for the construction of the other two new walls. Once the pegging out has been completed then Andy will ask Steven to 'make it right'. This offers Steven the latitude to construct the wall on his terms with his craftsmanship and experience guiding its completion. It's not a total hand-over because Andy has to choose the location of the stile entrance, and, in consultation with Steven, decide the position of the steps, and select the stones to occupy the crucial positions at the top of the wall either side of the entrance gap.

Andy will check these when they have been built and also the dimensions of the fold, especially the height. All the other features of the construction follow to the letter those that are traditionally in place in the surrounding walls. If there is any debate between waller and artist about any feature in the construction it is Andy who has the ultimate say about what the finished fold looks like.

Their professional relationship is fascinating to observe. There is a mutual respect for the other's contribution to such a work. Andy would not wish to enforce the notion of 'Artist' upon Steven's work, nor does Steven think of his craft as an art form or himself as an artist.

Andy could build the folds, as he has the skills to do so, but he has often acknowledged that they would be '....less powerful, less strong for having ignored the day to day craft of a lifetime's walling from a professional master craftsman'.

The most revealing comments from Andy concerns his fears about the influence he may have as artist upon the particular and idiosyncratic structural skills of the wallers employed to build the 'sculptures'.

'Art in the context of constructing the pens would be restricting. Any move towards the art form by the wallers would start to produce indulgences on their part which would begin to show in their work. Their focus would be different. The strength of these sculptures is that they have been built, without influence, in their traditional style by some of the best wallers around'.

In the broader artistic context the situation is not unlike the painting pupils who filled in the broad areas of Leonardo's design or the foundry craftsmen who produced the scaled up castings from Henry Moore's maquettes.[18] If the finished product is recognisably the work of its creator there is no problem. As soon as the accomplices begin to take artistic decisions of their own in preference to their employer, and it shows, then one imagines the terms of contract will quickly terminate.

In this present project, all the sheepfold exhibits are advertised as the work of Andy Goldsworthy when often he hasn't put a stone on them. They are definitely Andy Goldsworthy's sheepfolds not Steven Allen's. This doesn't affect Steven. He has no ego. He simply openly expresses his love of being involved with them. Only a handful of people know that Steven and his wallers have done the work on the pens. What is good is that their work and skill **are** always acknowledged by the artist, even if rarely so by the media and the public in general.

Fold at St Albans

---

[18] **Maquette:** A small, hand-sized prototype idea for a sculpture usually in clay, from which a larger scale version may be cast or constructed

The folds have been built on some unusual sites and there are two to which I particularly want to refer. The first was built as a temporary exhibit at the art gallery at the University of Hertfordshire. In order that the contextual reference of the original project could be reproduced accurately, the stone was selected by Steven from Cumbria, and transported south to the gallery.

Close to the entrance is a floor-to-ceiling windowed wall, which offers a permanent viewing access to the gallery inside. Andy Goldsworthy designed a fold which created the illusion of being built through this windowed wall by constructing half of it outside and up to the glass and butting up exactly the remaining half on the inside of the gallery.

The resulting effect was stunning! He brought together two wall construction methods, which though totally different in material, function and context, embraced each other as a work of art.

St Albans' fold rebuilt in Cumbria

Why this fold didn't remain a permanent exhibit at the gallery is a puzzle. What is interesting is that it was transported back to Cumbria, re-sited in the its familiar location at Barbon near Kirkby Lonsdale and remains Steven's favourite of the folds that he has built.

Mungrisdale Fold – detail

My favourite exhibit remains the field boulder fold at Mungrisdale. This for me is the ultimate treasure so far in the whole project. It draws together in time, every aspect of dry stone walling as a craft, with its development in this form as an art exhibit.

In the corner of a field rests a huge pile of stones. There is nothing extraordinary about this for the pile looks like the typical corner dumping spot for all the boulders and rubble that have been cleared from the enclosure.

Fold at Mungrisdale - exterior

It is only when you approach the pile and climb over it that you discover that in its core is an inner circular lining of constructed dry stone wall having a walk- into entrance and natural rough cobbled floor, making it the perfect protective sheepfold. It also has all the combined elements of technical strength of proportional and aesthetic beauty, which contrive to create the surprise of discovering it as a work of art.

An unprepossessing pile of stones, protecting a circular, hand built, retaining enclosure of stones, which in turn encloses and protects livestock, in the wider context of a typical, Cumbrian, community farming landscape. Breathtaking! This was the inspiration for the title of this book.

The American Connection

The mutual relationship of trust between artist and artisan developed further when Andy invited Steven to lead a small, highly skilled team of wallers to work on a project in New York State in America.

The commission was to design a sculpture in the form of 240 metres of dry stone wall, which would have 3 special additional stone features in its construction. The stone for the features was newly quarried, blue in colour and resembled slate. The rest of the stone was similar to that which Steven works with in his native Cumbria.

The three feature walls were to be built with the layers of stone tied and coursed vertically; within their belly at some point, a boulder of Casterton proportions, positioned between the footings and the line of coverstone and cope.

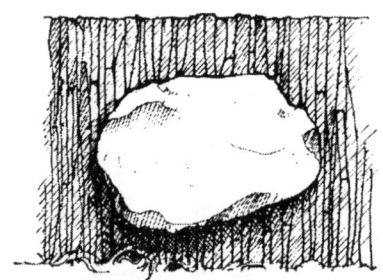

Boulder in Wall, U.S.A.

There were two main problems to overcome: firstly to find a way to course the stone vertically without the wall peeling away from itself; and secondly the arrangement of the base stones to support the stone in immovable, egg-cup fashion.

New York Project

Steven invited three other master craftsmen who wall from different bases in England to accompany him on this project. Two of them, Gordon Wilton from Derbyshire, and Bill Noble from West Yorkshire have also, subsequently been winners of the National Championship. Jason, Gordon's son, is also a keen and successful Grand Prix competitor. They are all outstanding wallers and together make a formidable team. Under Steven's supervision and Andy Goldsworthy's direction they completed the task in four weeks and also had time to prepare another project in a nearby sculpture park, which would be completed a year later.

The site of this second American exhibit follows closely the overgrown remains of an existing wall which at one time marked a boundary line set in a plantation of mainly deciduous trees before the trees were felled and the land cleared. This original wall went in a straight line and the edge of the existing plantation marked by it is also described by the line of trees that have grown from and through it. This wall ends at point about 60 metres above a small reservoir.

The stone for the wall almost certainly came from the land clearance and built by settlers from Northern England and Scotland.

It was Andy Goldsworthy's idea to create a new route for a new wall which 'flowed rather than cut' its way between the trunks of the trees at the edge of the plantation. The subjective choices of how the route made its natural, logical path between the trunks were limitless except that it had to 'flow'. Andy made sense of it by continually walking it, pegging it out, adjusting it until it felt right. The finished route snakes its way between the trees, is interrupted twice by roads and dives into the reservoir at its conclusion.

Walking Wall, U.S.A.

Some of the curves are so tight that it is almost impossible to stand inside them, whilst others open out into graceful arcs. Cheek ends stop the walls at the roads. Whereas the original wall stopped above the reservoir, the new wall descends and slips into it.

To create this illusion the reservoir had millions of gallons of water drained from it to allow the foot of the wall to be built before re-filling the reservoir to cover it. There are thoughts that next year the wall may re-emerge at the other side before taking a similar, climbing route. As there are few trees on the other side the wall will follow the undulating contours of the landscape.

Once the route was pegged the footings were dug out by hand, the copes and throughs laid out on wooden pallets and the rest of the stone placed in piles at convenient intervals along its length. When Steven and his team arrived in the autumn of 1997 they were able to begin work immediately.

As with all new walls it takes a little time to adjust to the condition and shape of the indigenous stone. The special feature of this wall is its continuous snaking curves. Once these conditions had been sorted the four walling and one labourer team would construct quickly to the highest technical standards.

A film records its construction. The camera follows the daily development between and around the trunks of the trees. Whilst I delighted in watching the wallers solve the idiosyncratic detail of these curves and their passage along the sloping ground, nothing prepared me for the final aerial shot, and the stunning effect of the whole line of wall as it snaked its route through the trees and slid quietly into the reservoir.

In the early period of the walled division of land, the actual routing of the wall would almost inevitably be measured and marked by the simple method of sighting strategic parts of the landscape and pacing lines between them. If an outcrop of rock or a boulder occupied a space somewhere within these lines then the walls would deviate accordingly to include or avoid them.

In designing his wall close to the site of the original, Andy Goldsworthy reflects his Cumbrian fold philosophy; it seems to be part of the same project. Once again he makes similar connections with the past. On this site the trees have matured as the wall has tumbled and 'gone to earth'. There is a visible timescape that marks its own tale of evolution. The new wall makes real associations with its historical, physical context and it harmonises with the immediate landscape because essentially it has grown sympathetically with it.

Steven loved working on this exhibit. It gave him the opportunity to create with his walling team a new, beautiful dry stone wall unlike any other in the world. It posed a special challenge of slope and curve along the entirety of its 360-metre length and it also had the familiar challenge of completion to meet an eight-week deadline.

In practical terms of course, like competition success, walling on these high profile projects is good publicity for future work. Steven has never turned down unusual commissions. He has a confidence in his own ability to solve highly technical challenges, which comes from his experience of problem solving on a daily basis. He also has a genuine desire to extend the boundaries of his craft.

Other Projects

Johnny Walker Whisky Bottle: Scotland

This was a commission by the manufacturers of the well-known Johnny Walker whisky to design and build a dry stone replica of one of their whisky bottles in the Scottish lowlands between Selkirk and Moffat.

Whisky Bottle

The site chosen was a remote and uninhabited glen. An enclosure of tumbled walls provided the stone ready to hand. The 'bottle', once constructed, would be photographed in its mountainous setting and the result used in a future advertising campaign.

Steven was assisted by Stephen Harrison, of Lawkland near Settle, a Master Craftsman and teacher of the walling craft. Between them they built a four-metre dry stone replica with its familiar square section, metal cap and angled label. The strange thing was that when complete there was nothing incongruous about it in the context of the wild and desolate landscape. It had a chimneystack silhouette that more than hinted of an earlier occupation.

However, once the photographic work had been completed the 'bottle' was demolished and the stone returned to its original tumble. One would have thought that it could have been reconstructed on or close to the site of the distillery as a permanent exhibit. As it is, the finished photograph tinted in the hues of the amber liquid makes a happy centrepiece above the mantel in the kitchen of Steven's home. This was just the kind of extraordinary project and challenge that Steven relishes.

Bungalow at Greystoke

Quite early on in Steven's professional career he was asked by a local builder if he was interested in taking on the building of the outer skin of stone for a bungalow which would be lined inside with modern materials such as breezeblock. Steven agreed and was immediately thrown into one of the most demanding projects of his early self-employment.

The reason for this was not the building work; it was technically straightforward. It was the preparation of the local Penrith redstone that took up a disproportionate amount of labour both physical and mental.

Bungalow at Greystoke

The owner insisted that the stone taken from local tumbled walls, carrying with it the weathering of centuries, be faced to reveal its original colour. The cost of purchasing newly quarried stone was prohibitive and so Steven embarked upon seven months of building, dominated by an excess of laborious cleaning and facing up each stone used. In addition to the work on the bungalow, a gateway, frontage and retaining walls completed the contract.

Most days Steven completed only 1 square metre of walling. It says a lot about his mental strength and qualities of focus that he was able to shut out the mind-torturing task of re-facing each stone and simply got on with it.

It also revealed his propensity to adapt his dry stone skills to the needs of general building work and masonry. Once he had completed this contract, he felt he could tackle most things. There are many professional wallers throughout Britain who will devote their working lives only to the repair and construction of field and garden walls.

Steven's love of working in stone generally has led him to explore the huge variety of methods of constructing with it. He is inquisitive and loves a new challenge, and it is this open-mindedness which has led him to a wide variety of work and, as a result, extended his range of skills.

He often finds that he is sometimes asked to take over work contracted to others who have not produced results to the required standard, failed to meet contract deadlines, or simply worked at such a slow rate that the resulting cost of a job becomes outrageously prohibitive.

Such a situation led him to the property of Member of Parliament David Maclean. Despite the open landscape that spread away from his beautiful house, he wanted a dry stone wall built around its perimeter and, in some parts, high enough to secure proper privacy from the walkers who use the public footpath, which runs along one side of the property.

Cumbria Garden

He also wanted the existing landscape garden to have areas of retaining, decorative walling involving arches, curves and corners around certain features such as pools and raised borders.

The available stone, moved from a derelict building within the grounds, was rough, cobbled, dense, granite. I spent a couple of sessions with Steven on this site sorting stone with him in fairly wet conditions and became aware of the vulnerability of the singular waller in isolated landscapes, working with such material and of the risks involved.

Some of the foundation stones were almost immovable, and the throughs in general impossible to lift without assistance. And yet, there is no question of any one else being available in such circumstances unless there is enough extra money in the contract for Steven to be able to share the work. He is left to find ways and means to resolve such problems alone.

As I have mentioned before, Steven is blessed with a strong and resilient body but some of the lifting and carrying tasks he takes on alone defy belief.

Garden perimeter wall

The main perimeter wall was built in a random style with a combination of granite boulder shapes and a number of tying 'flatties'. In some places it is over two metres high and has a steep bank falling away from it increasing the illusion of its 'impregnable' height. The stone was laid out along the length of the dug out foundation trench, and, once a section of wall was completed, the remaining stone was shifted along to add to the next building pile - and so on.

Once the building pile becomes a real pile of rubble, only fit for filling, then more stone is collected and laid out. Steven kept coming back to this wall in stages as he completed a section and the owner organised a local contractor to shift and replenish the stone.

This kind of contract is also useful because the owner of the property, a public figure, can pass on Steven's reputation to others who have the wealth to buy his skills for domestic walls to their property but also to bigger contractual work that may be conducted in the area.

Long roadside wall

Such a contract came Steven's way almost by default. He had put in a tender for 500 metres of wall that bordered a new highway, which was to be constructed to by-pass the small town of Haltwhistle. It had been turned down in favour of another waller. Sometime later, after the wall had been started, the original contract was terminated Steven was contacted by the developer asking him if he would build the wall. Steven re-negotiated the contract, sorted out his existing walling commitments and took it on.

Long roadside wall

Ironically this wall was probably the first of this scale to be built in the area since the great enclosure period of the late eighteenth and early nineteenth century. Just north of this wall also stands some of the finest remains of Hadrian's wall from the time of the Roman occupation. When I visited this site I was also aware of the hand of time and history repeating itself and Steven's link with it.

This contract marked another development in Steven's career; that of man-management. In order to meet the contractual deadline, Steven had to employ a team of eight wallers to guarantee its scheduled completion. Three of the wallers he knew personally and their work standards, others he had to engage by interview, and hearsay. It turned out that those in the latter group worked to varying standards, and the results are visible in the wall.

Roadside wall with stoup

There are some wallers who will throw a wall up in order to maximise their rate of pay ignoring the basic building criteria of style and strength in favour of speed. It also encourages ill feeling between those wallers who work hard to complete a wall to the required high standards and those who are more focused on monetary reward. It dissipates and devalues the essential element of teamwork.

In the end the responsibility for the finished wall is Steven's, and it will inevitably have his signature upon it. Whilst he would not expect others to match his standard of competitive walling he has the right to expect a recognisably sound walling technique from the professionals he employs.

In this instance he was so committed to a critical deadline that once into the walling timetable he could not afford to lose those working for him as replacements would be difficult to find at short notice. Some wallers, after a few days of walling, refused to work with others whose skills and work rate differed radically from their own.

For the first time, Steven was faced with the tricky task of sorting such problems out so that the momentum of the work was not disturbed and the contract completed on time to the required standard. In the end Steven spent a disproportionate amount of extra, personal time tidying up sections of inferior work. The resulting wall completed on time - a remarkable achievement in itself - was solidly built, and looks well.

The main lesson Steven learned on this contract was the absolute need for him to be able to guarantee the right standards and attitude from the wallers he employs and that the only way to achieve this is to see them at work or work with them.

Which is another reason why walling competitions in general bring you into contact with good wallers who not only have a high degree of skill but who take pride in their work and are not afraid of public scrutiny. It is from such a pool that Steven has employed wallers to work with him on such projects as the Andy Goldsworthy sheepfolds and the American sculptures, and why they have all shared the pleasure and challenges of new styles of testing work - and each other's company.

Oasis

Such a waller, who has only recently become a full-time professional is George Allonby, who came up through the ranks to win the National Amateur Championship in 1997. George lives close-by in Penrith and now does a lot of contract work with Steven.

Entrance at Oasis

The small project at the Oasis Country Park consisted of building two architect-designed dry stone bases to support display signs either side of the entrance. Illumination devices and a small fountain would require wiring ducts and piping to be hidden within the structures.

The stone had to be collected from a general tip of quality red Penrith stone within the Oasis complex, and the filling touched with cement without losing the exterior dry stone effect.

Given Steven's experience with the Bungalow a few years previously this exercise was completed to the highest possible standard with the minimum of difficulty. He could do this sort of work in his sleep.

Rusland wall avenue of beeches

The area of Rusland, south of Lake Windermere became the centre of a huge swell of local protest concerning the felling of a number of mature beech trees on the side of a minor road. The public reasons for felling was because experts discovered advanced stages of disease at the hearts of the trunks of the beeches and there would soon be a real risk of their cracking or falling in a heavy wind.

As there were no outwardly visible signs of the problems described by the scientists, the locals thought it pertinent to challenge the authorisation to fell the suspect trees arguing also that in all probability there was a hidden political agenda, perhaps even the suspicion of a possible new building or road development.

The trees were eventually felled, and found to be in the condition the arboriculturists had described. However it was decided to repair and rebuild to the highest standards required by the National Park landscape officers, the existing crumbling dry stone wall and to fill in the new gaps along the road. The resulting work coincidentally was contracted to Steven, who, with George duly completed the wall to the required standard.

Whilst the local residents miss the trees, they are happy with the replacement walls. There are few occasions when dry stone walls become the centre of controversy and this was an occasion when their construction, in the aftermath of landscape mutilation, helped to soothe local discontents.

Sculpture at Heathrow

Shortly after returning from the 'walking wall' project in America Andy Goldsworthy asked Steven to work with him on a sculpture, which was sited both inside and outside the British Airways offices, just beyond the perimeter of the Heathrow complex close to the village of Harmondsworth. Coincidentally this was Andy's first piece of commissioned work for a private body.

Inside the building close to the entrance, Andy had already constructed one of his familiar dry stone 'cones'. It was the companion piece that was to be constructed outside within a newly designed landscape.

The structure is a rectangular cuboid 3 metres long, 2.7 metres high and 1.5 metres thick built with the same, fairly narrow slabs of Cumbrian stone. On the side facing the building, the cone shape is scooped out. It is as though the cone has been removed leaving half its mould standing; a bit like a shell without its nut. Whilst Andy completed the outline of the cone Steven built the containing cuboid around it. The curved 'roof' of the hemi-cone provided the main technical challenge. The resulting association between the interior and exterior exhibit is simple and powerful.

Chelsea Flower Show

In late April and early May, Steven was commissioned by The Daily Telegraph to build a dry stone construction designed by Sarah Raven for the Chelsea Flower Show. With Bill Noble and Kevin Alderson assisting the resulting exhibit contained a series of small enclosures filled with a contrived tangle of mature plants. It was built to an exact specification and provided Steven with another collaborative experience with an artist. It took only four days to build.

Apart from defining the perimeter of the exhibit, the wall retained a border, marked an entrance, tumbled convincingly in places, supported a turfed cope, edged a pond, bordered steps and a rough walk of stepping stones. Apart from it being technically strong it had to appear 'roughly thrown together' and form an unobtrusive backcloth for the plants, which were draped like discarded clothes over and around it.

It was fascinating to see the completed wall skeleton before and after planting and it is these two states that I represent in the drawings below.

Chelsea Flower Show – exhibit under construction

Chelsea Flower Show exhibit on completion of planting

CHAPTER THREE

Competitive walling

Kilnsey Open – a personal recollection

The early morning mist hangs, nudged by the queue of horse-boxes, Land Rovers, vans, trailers and tractors, shuffling their way in single file to the competitor's entrance The road cuts into the limestone cliff which hovers like a billowing cloud above it. Kilnsey Crag, the infamous Yorkshire overhang, presides, scowling over the temporary canvas showground.

During the hour or so in which the competitors assemble and register, there is much checking of the arrivals - and the opposition. Some greet each other as friends catching up on their year in a couple of sentences. Others 'rib' at changes in paunch and hair loss. Some bring wives and children and ignore them until the break for lunch and the journey home. Some search beyond their laughter towards their sworn rivals.

All of them look at the wall, checking the measured stints for the advantage of the quality and variety of stone and especially for the one they hope they won't draw. Two of the finest wallers in the land happen to be amongst the competitors, circling each other without acknowledgement, each surrounded by his inner clutch of worshippers.

The wall to be demolished and re-assembled is built with random chunks of limestone. The rules for this competition are different from the norm in that hammers, lines[19] and pins[20] cannot be used.

---

[19] **Lines:** Traditional builder's line attached to metal pegs which are used to keep the courses of stone both level and sloping inward evenly (see "batter"). These follow the line of the walling pins or batter frame.

[20] **Pins:** Steel rods driven into the ground at the footings which mark the vertical cross-section of the wall from the footing to the cope to which the lines are attached and shifted with the lifting of each course on either side of the wall

This means that both straightness in coursing[21] and batter are achieved by eye alone. For some this will be an irreconcilable handicap.

As the time for balloting approaches the competitors hump their picks and shovels to the roughly pegged line, running parallel to the wall, beyond which the public may not enter. Some, superstitiously, place their tools in front of the stretch of wall they judge to have the worst stone.

The event is opened officially by the Chairman of the local association as he calls the wallers to his tent and, because this is the centenary show, a group photograph is taken. He welcomes, checks the names of those he thinks have registered and then offers a tray of folded slips of paper that conceal the numbers of each stint.

The rules are simple. Within the allotted time of seven and a half hours, 3 metres of wall have to be demolished - including the removal of the footings - and rebuilt, and there is no compulsory lunch break.

Once the competitors have identified their stint number, they move to it, curse or smile at their fortune, and check out who occupies the adjacent stints. It so happens on this day that I have drawn a stint between the best and the second best wallers in the land.

Whilst I stand little chance of chasing any prize my main preoccupation will be to build a wall which is straight in cope and batter and which, at the end of the day, does not catch the eye for the wrong reasons. I have ambivalent feelings about being wedged between these two stars. On the one hand any poor work on my part will be mercilessly visible but on the other, if I stay relaxed, I can keep half an eye on their progress, learn from it and use that knowledge to my advantage. At least that is the plot in theory.

The starting whistle is blown. The two either side of me lift the stones at the ends of their stint first to establish clearly the

---

[21] **Coursing:** The method of keeping each layer of similar depth stone to the same level as the wall "lifts"

demarcation line between us, hit their wall like wild things and change into a whirlwind of demolition. One, dripping sweat, sorts his stone into sizes close to the wall. His rival tosses his, apparently without sorting, into one, spreading, reachable pile.

The one on my left digs out the footings and softens the earth with his pick before anyone else. He stops, swallows the contents of a bottle of Lucozade and glances at the progress of his rival who is only just collecting his pick. I am in a similar position and we prepare our footings together. As we do this, a steward walks along the wall unwinding a line. This is the only aid we will have before it will re-appear and be lifted to mark the top of the cope at the finish. For the rest of the day the constructing of the wall is in our hands.

The stone is irregularly random, triangular, and dense. It would be impossible to crack it with a hammer had the rules allowed its use. There are two sets of throughs at knee and elbow level and the overall height of the finished wall is about 1.5 metres.

Each competitor is responsible for tying into the left of his neighbour's stint, although I find that this is taken care of at each end of my wall by my competitors, probably to ensure that I include nothing in my stint that will encroach upon the general look of their work.

Sometimes just one stone, wrongly placed can change the effect of the whole. They were working overtime on the edges of their wall to ensure that this didn't happen. I was grateful. It meant that the edges of my wall matched the standard of theirs - they had built it!

On my left, the wall grew rapidly beyond my pace, on my right it matched my own, and whilst this remained so I felt confident that my progress would not be too obtrusive. I was conscious of the singular judge spending a lot of time looking at every detail of the stints either side of me giving mine no more than a cursory glance.

It seemed the easy option to simply replace the footings as they'd been removed. In my stint there were a couple of stones that had been placed in a trace position. I corrected this and added another stone from my stock. This was to cost me at the end of the day. Once I'd completed the first course I began to feel more confident with the stone, and the early tension I felt putting it together eased somewhat.

Either side of me little was happening and so I continued through the first lift focused on my own building. By the time I had placed the first throughs in place and was taking my lunch break I was feeling confident enough to predict that I would finish my stint and that I would not make a fool of myself.

One of the interesting results of not using pins and lines to aid the straightness of the coursing and the batter of the wall, is that there is a greater sense of shared responsibility amongst the wallers to make sure the continuous line of the wall stays true; an overall will to get the whole thing straight. When pins and lines are used in other competitions there are often a series of zigzag ripples where individual stints have been constructed to the positioning of the pins rather than to the whole length of the wall.

At this lunch break, my adjacent adversaries are talking to each other but don't appear to be looking at their work. Other wallers join them severally checking their stints for them and some make openly provocative comparisons. There are no specific rules concerning the way the wall is to be built. It is up to the individual, local style and strength being the only criteria.

It is also obvious that they are rebuilding it in subtly different ways. One is keeping strictly to a random style whilst the other is following his renowned coursing. It will remain to be seen which of the two especially catches the judge's eye.

One of them joins his family and sits amongst an enormous spread of picnic, the other strides to his jeep and closes the door on himself, his food and his girlfriend and will remain so until he is ready to rejoin the fray.

About them, the rest of the show is changing its focus according to whatever new event or category of beast is timetabled. Fortunately, because the walling event follows one side of the main highway the crowds continue to stop and look at the progress. Even during this break both sides of the wall are lined with onlookers, and some, more inquisitive than others, are keen to make real contact with the wallers and stop to talk to them. Others, not competing but obviously wallers, offer their opinions anyway. Standing back from things allows competitors and spectators to take stock.

I sit with an old waller who now is one of the chief judges on the Grand Prix circuit. He'd won this competition five times. He is looking at the walls either side of my stint and is finding it difficult to decide who is ahead at this point and knows that the final judgement will be contentious. He is glad this responsibility is not his today.

It is not long before the spectators drift beyond the ropes and the wallers start shifting stone once more. During this second lift the wall seems to grow at the same pace as the competitors catch up with each other and feel the common focus towards the finish. There is nothing more threatening to a waller who finds himself falling behind those on either side. Today this is not happening, even with me.

At the end of the second lift and after the setting of the second throughs the line of the whole wall is as straight as a die. All the action has become perpendicular. The numbers of spectators have increased appreciably and there is now a buzz of expectancy in this final stage of reconstruction.

Either side of me the final lift has moved quickly and the one on my left has already set the ends of his cope in place, and these are being used by the stewards as one of the sets of guides to support the final cope line for the length of the wall.

I'm struggling. I usually do at the finish. I've plenty of time left to complete but I'm now aware of most of the other wallers putting on their cope stones whilst I'm still finishing the final lift. Being aware of this is sufficient to further slow my progress. By the time I'm ready to begin my cope everyone else has finished. The two rivals either side of me make occasional adjustments to their copestones adding or subtracting small wedges to improve alignment or tightness.

One stands back and takes off his gloves - a sure sign that he has finished - whilst the other continues to make adjustments. I'm glad of his company because otherwise I would be the only waller still constructing. As it is I spend the final ten minutes finishing alone.

By the time I've cleared my stint the whistle blows signalling the official end to the competition.

The large crowds of spectators who have witnessed this final flurry re-group to form a loose semicircle around the presentation area where a table is being erected and trophies displayed by officials.

There are prizes for younger wallers under the age of twenty-six years and there are prizes in the main event right down to the eighth position. However most attention is focused upon the first two positions to see who has triumphed in this particular arena. The general opinion is divided almost equally between the arch rivals, although most share the view that it will be up to the judge's preference for either a coursed or a random style in this instance.

The runner up is announced, and the competitor concerned moves to his rival to shake his hand before he turns to collect his second prize. His taut, unsmiling grin and his lowered head catches his disappointment. As he moves back to his family he bends further, applauding the cheer that accompanies the announcement of the winner.

Once the prize giving is complete there is much to-ing and fro-ing between the stints either side of my own, much pointing and tapping with boots at questionable areas of construction, fingers testing small filling stones for looseness, and whispered asides.

And me? I came in the 'also ran' pack. However it was the most enjoyable competitive walling I had completed during the year. Working between such marvellous craftsmen had been a joy, and I learned much.

The DSWA Grand Prix

I mentioned Steven's involvement and successes with the Grand Prix competition in my introduction to this book. He was the first to hold the winner's trophy and he has held it several times since. It has been one of the main focal points of his competitive walling, and he champions its existence.

The Grand Prix was first established as a competition in 1991. Its purpose was to encourage wallers from all over Britain to meet and compete in a variety of locations having differing walling conditions. The challenge was to see how wallers brought up on the indigenous material of their homeland would apply and adapt their local constructing skills to a selection of different styles. Hopefully there would be enough wallers willing to travel to compete in different locations and at the end of the series the waller with the highest total of marks could be crowned British Champion. These competitions would take place between late spring and early autumn.

At the present time there are seven competitions and this year (1997) they have been held in Yorkshire, Lancashire, the Cotswolds, Wales and Scotland. Wherever possible the DSWA will encourage local branches to organise their competitions within the context of a wider public event such as an agricultural show or a gala day held in the grounds of a Great House. In this way it is hoped to attract a wider audience and encourage interested spectators to join the association.

The rules for each competition remain more or less the same the only variance being the dimensions and style of the wall. When the wallers gather in the early morning they will expect to find that the wall which is to be rebuilt, has been pegged out on both sides of the wall into the measured stints according to the singles or doubles categories of professional, amateur, novice, or junior; their length matching the expected experience in each category.

Five or six metres from the wall on both sides will be roped off along its length to distinguish the safe working area of the wallers from the spectators. Sometimes, if a particular stint is obviously short of stone, a replacement pile will be placed in that stint.

There will usually be a small marquee from which the local group will administer the event and the DSWA may also organise another covered area to display material and publications produced by the association. Once the registration of the wallers has been completed then the lot numbers corresponding to the stints are drawn. The wallers then take their tools to their stint and check to see that there has been no oversight in conditions in their stint which may be a disadvantage to them.

Any special conditions such as a large boulder, or a particular lack of a particular sized stone will have been noted by the judges and accommodated in their final judgements.

Judges

These are Master Craftsmen who themselves have sometimes - though not always - had a successful competitive career. Their main task is to observe the stripping down and to award marks at each stage of the reassembly of the wall. These are usually for the footings (foundations), the first lift,[22] the throughs (where applicable), the second lift,[23] coverbands[24] (where applicable) and cope, batter and straightness. By the time the wall is completed therefore a lot of the marks are hidden in its earlier construction.

There is also a category of marking which includes the general organisation and safety conditions that are observed in each stint. Whilst the techniques employed in the placing of the filling can be marked in each lift it may be included in this general category. In the event of a tight finish between competitors it is from this category that the judges can make their final judgements in favour of one over another.

Whatever happens, theirs is an exacting and often thankless task as they follow the progress of all the wallers throughout their reconstruction of the wall and beyond its completion.

The whistle blows.

1: STRIPPING OUT[25]

No two competitors will "pull" down a wall in the same way. Some will lay the stone out neatly on both sides, a course at a time in the hope that they will rebuild it approximately as it has been previously constructed. Others will tear it apart and throw the stone into a mixed pile of fillings their only concession to any semblance of visible order being their separation from the pile of throughs and

---

[22] **First lift:** The construction of the wall up to the first line of "throughs"
[23] **Second lift:** The construction of the wall up to the second line of throughs or to the copes
[24] **Coverbands:** Flat slabs of stone which tie and cover the top of the wall below the cope
[25] **Stripping out:** Pulling down a wall, stone by stone, and laying it out on both sides

copes. Others will grade the size of stone as they remove it according to its proximity to the wall to be rebuilt, the largest closest to hand and the smallest further away with the throughs and cope in their respective places in the general scheme of things. Usually the fillings will be organised into reachable piles. There is no official way of doing this. It is up to each individual to do it the way that works for them.

Stripping out

In any case the wall has to be removed to the last stone of the foundations unless there is a boulder which is obviously too huge to dig out. In such an instance this is usually confirmed by the judges and will be left in place. In the end the channel for the new wall is dug out so that all that remains is earth. At this stage the main tools which are used are picks and shovels.

Once all the competitors have dug out their footings then a line is established - by the stewards under the direction of the judges - along the length of the competition wall on one side. This affords a short pause in the proceedings while this is fixed. Once this is done then the construction of the wall begins.

2: THE WALL AND ITS DIMENSIONS

This varies according to the indigenous regional style. Some years, whilst the basic dimensions of the cross-section change only slightly, the use of throughs, coverbands and copes vary considerably. As mentioned at the beginning of chapter 4 the basic dry stone construction consists of two walls wider at the base than the top, packed with a filling of small stones and tied at intervals with throughs and covered with a continuous line of coping stones.

In the Cotswolds there are few stones large enough to use as throughs. There are hardly enough large stones to place in the footings. The wallers therefore place even greater importance upon the quantity and placing of the filling. In the competition this is reflected in the mark allocation.

In the Cumbrian wall there are two lines of flat slabbed sandstone throughs which sandwich courses of rounded, granite stones.

At the Pentland site near Edinburgh the random wall was short in height with a single through, topped by a coverband of huge, dense, footing - like boulders which were also covered with a smaller stone cope.

In the elevated regions of Rochdale the stone was already laid out for the construction of a new wall. The competitors therefore had to sort out their own throughs, copes and gradation of stone and build to the more general 'one through and random cope' style of the region.

And many copes are different, tying the two outer walls as they do in a random, castellated, straight, leaning or flat style. It just depends what method has proved over the centuries to be traditional for the region. These differing styles identify location also and thankfully are now being keenly demanded in all new walls to preserve that identity.

So ... rarely are two styles the same, nor is the type or shape of stone. This is the exacting and exciting challenge that faces the regular competitor. Obviously competitors on their local patch should be at an advantage, but the overall challenge is to find those who adapt best to new conditions and who apply best the fundamental principals of good walling technique.

3: FOOTINGS

Once all the loose stones have been removed and a rough tilth of soil created (with pick and shovel) then the foundation stones are eased into place. They should - as far as possible - be set into the wall rather than along it, but obviously if a huge boulder sits happily lengthways then it is used in this way.

Footings

Whatever happens the stones should be dug in (often using a hammer) eliminating any trace of wobble, packed tightly with fillings and wedges, edged as neatly as possible with the prescribed line marking the foundation width of the wall, and be as level as the randomness of the stone allows. A level, straight, tight, secure foundation provides a solid base upon which the rest of the wall is built. Each waller is responsible in the construction for tying into the wall on their left.

4: FIRST LIFT

Where the wooden pegs mark the end of each stint, some competitors will hammer in pins at the edge of the footings at the recommended angle, which determines the batter of the wall. The distance from the footings to the cope is marked and the pins tied at this point to establish the width of the top of the wall under the cope.

The first stone is placed and a line secured at this height across the width of the pins. The stones are then placed along the line though not necessarily following it exactly like a line of bricks. The line is in place to remind the waller of the sense of an imagined horizontal that should be followed. With a random wall this line is often broken but it still acts as a guide.

Ideally, the larger and rougher stones are usually placed at the bottom of the wall and they reduce in size as the wall lifts. Wallers tend to course with stones of similar depth where possible and keep each side lifting at the same level.

First lift

Once the next course stone height is established the line is lifted to it, and so it continues upwards.

In the day-to-day reality of walling, away from the competition arena, pins and lines will often not be continually used. But rarely - save in competitions like the ones at Malham and Kilnsey (see beginning of this chapter) will competitors wall without them, there being enough time to make good use of them in the general scheme of things. If the lines are followed then the straightness and batter should take care of themselves.

Stones, where possible, should sit comfortably across the joint of the stones below, fit tightly next to each other, be set as near to horizontal as the batter allows, wedged, supported and packed with filling to prevent them from moving once they have been set, and be arranged to lie 'inside out and outside in'[26] and not 'traced'[27]. There is nothing unusual about this walling practice which produces a competitive championship wall. It is just by observing these basic principles to the letter with each stone that such a wall will be built.

---

[26] **Inside out and outside in:** Making sure that a rectangular stone is set with its smallest end face to the front of the wall, its length set into the centre of the wall to produce added strength

[27] **Trace walling:** Placing long stones along the face rather than into the hearting of the wall (as above). This is generally regarded as being inferior walling technique and creates weaker structures.

The judges will be looking for fault lines or cracks up the wall created by joints being placed directly on top of each other and how stones are chosen to be placed in the general scheme of things. Strength and style dominates their marking criteria.

Most wallers will hope to finish this first lift by the compulsory half hour lunch break in the middle of the day and will hopefully have set their first throughs in position. Their thinking is that if this is achieved then they are ahead of rather than chasing the allocated time for completion. It is a psychological bonus to be able to settle down to some nourishment and rest with the wall already at this stage of its construction.

5: THROUGHS

In some instances these will have to be set so that each end matches the line of the face and batter of the wall (as in the Grand Prix in Wales this year) but usually they protrude on each side and become a visible feature along the length of the wall. The judges will be looking particularly at the way the through ties the courses below and above it, how it is supported by filling underneath, its stability (no rocking), the way it is set squarely across the wall, the way in which the courses fit neatly and tightly up to it on all four edges and continue in the same way along the line of the wall to the next throughs.

Throughs

The same procedure is followed if there is a second line of throughs above the next lift.

6: THE SECOND LIFT

This is the part of the wall that catches the eye first. It is built in the same way as the first lift only the stones are much smaller and can be assembled more tightly together, creating a finish which usually looks much neater than at the foot of the wall. Straight lines of coursing at the top tends to be the popular way of finishing.

Once the courses have reached the prescribed level then the wall is prepared to receive the coping stones which span its width. Most wallers will give extra attention to this so that the final line of walling is as horizontal as possible. If the height of the coping stones are the same then simply by laying them on top of an already level wall will ensure that it follows the same line.

Sometimes the wall is covered with flat slabs of 'coverband' which are butted up to each other in a continuous line where possible, and the coping stones sit on this. Providing the coverband does what it should do and cover the width of the wall the coping stones don't have to, providing they are of the correct height and weight.

7: COPING STONES

These span and sit on top of the wall. Whatever the specific style of the region they must be tightly butted up to each other. Where this is not possible then wedges of stone are used and jammed into the cracks between the stones.

Coping stones

Where an unbroken line at the top of the cope is required then for those stones that don't quite reach it added packing is placed underneath them. It's all fairly logical.

Most wallers will set coping stones at each end of their stint and strike a line between them passing over the uppermost centre point and secured beneath.

8: THE FINISH

Once the wall is complete the site has to be tidied. It is for most people the magic moment of the day when all the paraphernalia of construction - pins, lines, picks, shovels, hammers and buckets are removed from the wall and it is seen for the first time in its finished, uncluttered, pristine state. A brand new wall.

The Finish

The whistle is blown for the last time, all work ceases and the marks for the day are collated.

The Grand Prix tends to attract the attention of some of the best wallers in the land who want to test their skills in the hotbed of competition. The title of British Champion is coveted. It is in this environment that Steven has developed his craft, and enjoyed some memorable victories. It is also an arena in which wallers meet, exchanging ideas, styles and methods, and share and compare conditions of working practice, not to mention the striking up of friendships. By its very existence it promotes the work of the DSWA and, when the best wallers are competing, provides an instant showcase of the work of master craftsmen.

CHAPTER FOUR

Master Craftsman Certificate

Throughout this book I have referred to Steven and one or two of his working colleagues as being master craftsmen. Whilst this is a generic term which refers to the highest possible standards of accomplishment, it also refers to the extraordinary variety of walling tasks which the Master craftsman is expected to be able to construct in dry stone.

The DSWA of Great Britain is the only recognised organisation that trains, prepares and examines candidates through its Craftsman Certification scheme. The certificate is designed to test and reflect the highest possible standards of craftsmanship.

There will be those who have not been examined from a previous generation of wallers but who are universally recognised as being 'Master Craftsmen' and who set the standards for the current test. This, nationally recognised accreditation, is the one to which all employers can refer and rely. Too often in this present climate extremely poor work is in evidence, especially in huge lengths of wall that follow the edges of new, public highways.

These walls may look 'pretty' to the untrained eye but the evidence of poor walling technique soon reveals itself in the very short time it takes for countless gaps to appear where the untouched wall has begun to tumble.

The certification scheme should help to eliminate the 'cowboy' wallers and their shoddy work and protect the reputation of the craft and those top craftsmen who work hard to sustain that reputation. Unfortunately there are as many unscrupulous contractors as there are cowboy wallers cutting costs to line their pockets.

The tests are available to all individuals whether or not they are members of the Association and are graded at *Initial, Intermediate and Advanced levels* leading up to the highest grade of *Master Craftsman.* It is not my intention simply to replicate the details of the scheme as a comprehensive booklet explaining the details of the training, how the tests can be arranged, and the exact requirements for each stage of the tests can be obtained from the DSWA.

I want to create a general picture of the essential walling features that the Master Craftsman should be able to build to the highest possible standard. Having said that, it is inevitable that there will be some features that the individual will feel more comfortable building than others. Steven rarely will be asked to build arches or hog-holes[28] for instance, whereas corners, cheek ends, curves, stiles, sloped walls, he will encounter on a very regular basis.

1: **Basic wall structure.**

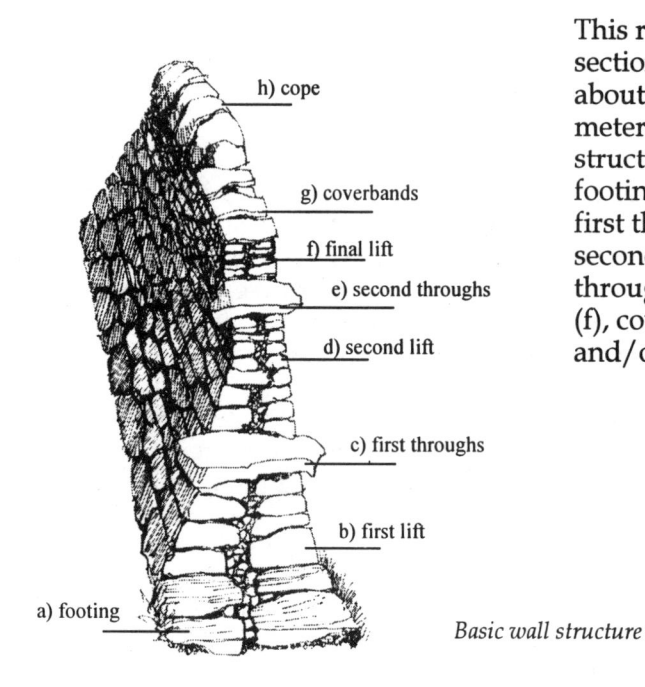

This refers to a straight section of wall usually about 4ft 6ins (1.3 meters) in height. The structure will include a footing (a), first lift (b), first throughs (c), second lift (d), second throughs (e), final lift (f), coverbands (g) and/or cope (h).

Basic wall structure

---

[28] **Hog(g) hole:** A specially constructed hole at the foot of a wall designed to let sheep pass through.

The basic wall design consists of two stone-width sides which taper inwards from the base, filled between with small stones (filling or hearting) and tied at regular intervals by throughstones which span them often emerging on either side.

The base is approximately twice the width of the wall at the top course supporting the coverband or cope. The sides of the wall will be straight along its length and perfectly flat along its angle. This angled face on each side of the wall is referred to as the *batter*.

As the stones are placed along the two sides of the wall, if they have any length they will be set into rather than along it. As each course lifts the wall from its base the joints between each stone are 'covered' by the stone above it thus 'tying' them more securely together. They will also, where possible, be set sloping slightly downwards from the centre of the wall so that rain drains off rather than into the hearting.

Footing

This refers to the foundation stones, which are set into a shallow trench dug into the earth, wide enough to accommodate the two sets of stones and filling which form the base for the rest of the 'double' wall.

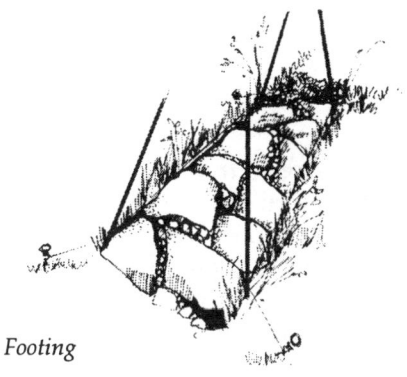

Footing

Where possible the earth should be cleared of stones so that a soft level tilth above the firm earth bed prevails. The stones should be the biggest (and often the ugliest) in the pile. If they have a flat side and a rough side the rough side can be 'dug into' the earth leaving the flat side uppermost to offer a true and level start for the next course. If the stones have any length they should be set length inwards. This is the general principal for setting stone in the rest of the wall. Each stone should be as tight as possible next to each other and the gaps between the stones filled with hearting.

Many wallers 'walk' along the footings testing each stone to see if they move at all, and packing or re-setting if necessary. Perhaps this is where the term 'rock solid' comes from.

First lift

First lift

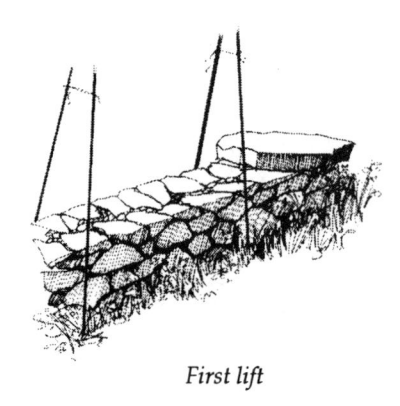

This refers to the first few courses of the wall that develop above the footings and go to the first set of throughs. This usually occurs at about knee height and the wall is levelled.

The batter of the wall is established and maintained by a batter frame (or pins) being set at each end to the batter, and lines between these set taut to guide each course as the wall develops upwards.

The filling is packed tightly between the stones in the centre of the wall and each facing stone placed as tightly as possible to those next to it.

Throughs

These are stones that are long enough to span (and often pass beyond) the width of the wall, and make good contact with the two sides. They are set at regular intervals of about a metre apart, and are designed to prevent the two sides of the wall from peeling apart.

Throughs

It is vital that the filling underneath them is packed tightly up to the through as any hollowness could result in fracturing of the through because of the weight of the stone above it. The throughs, where possible, are placed in such a way on both sides so that their edge does not coincide with the joint of the course below.

In some areas such as the Cotswolds throughs are difficult to find and so extra care and a slightly different technique is needed to lay the largest stones available between the sides and with the placing of the filling

Second lift and second throughs, final lift

This process follows exactly that described above. The stones will become progressively smaller as the wall grows. The final 'lift' of the wall takes it to a point that becomes the support for the coverbands or the copestones (cope). It is important to continue to take special care to maintain tightness of joints and level layers of stone. If this is neat and level then it is properly prepared for what sits upon it.

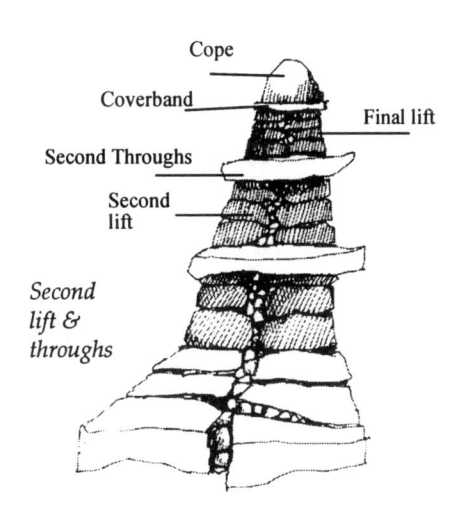

Coverbands and Copes

These span, act as throughs, and with their combined weight, tie the top of the wall. Coverbands are laid flat across the two sides and are usually butted up to each other in a continuous line. They also cover, and protect the filling. Sometimes coverbands support a cope. Copestones can be set vertically, angled, castled, randomly shaped or cut to a specific design. Even when the copes are random in size it is usual to strike a line to create a level top. This line especially is the one that always catches the eye.

2: Retaining

This is a wall that is built against and supports a banking of earth. Sometimes the wall will emerge from it at the top for one or two courses. The front face of the wall is completed to the same specification as a two-sided wall but behind this single face is an extra filling of rocks, which replaces the volume of rock required to complete the 'missing' face. It is also similarly tied with throughs that are set horizontally into the earth it retains.

Retaining wall

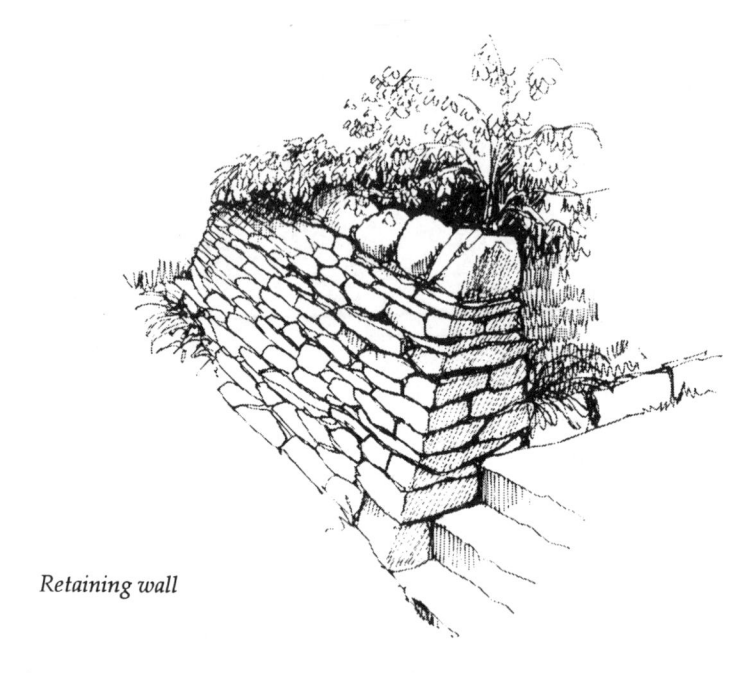

The width of the wall depends upon the volume and pressure of earth that is piled against it. In the preparations the earth must be cut back to a distance of not less than half the height of the finished wall. This cutting back may be extended if the earth is loose and liable to slumping.

The copestones should be as substantial as possible particularly if the wall is to be backfilled at the top. Whilst a dry stone wall is usually free draining where the backfilling is compacted because of the materials used then weep holes, (drainage channels or drains at the footings in extreme cases) need to be constructed for drainage at 2-3 metre intervals.

3: Steep slope

There are thousands of metres of dry stone walls that climb the steepest slopes of the British uplands. And these are the walls that seem to defy gravity as they cling precariously to the hillsides. They stand as a testament to the skills and physical endurance of generations of wallers responsible for their construction and maintenance.

Whilst their basic structural design remains the same as a wall on level ground the process of building, because of the extra problem of the slope, is different.

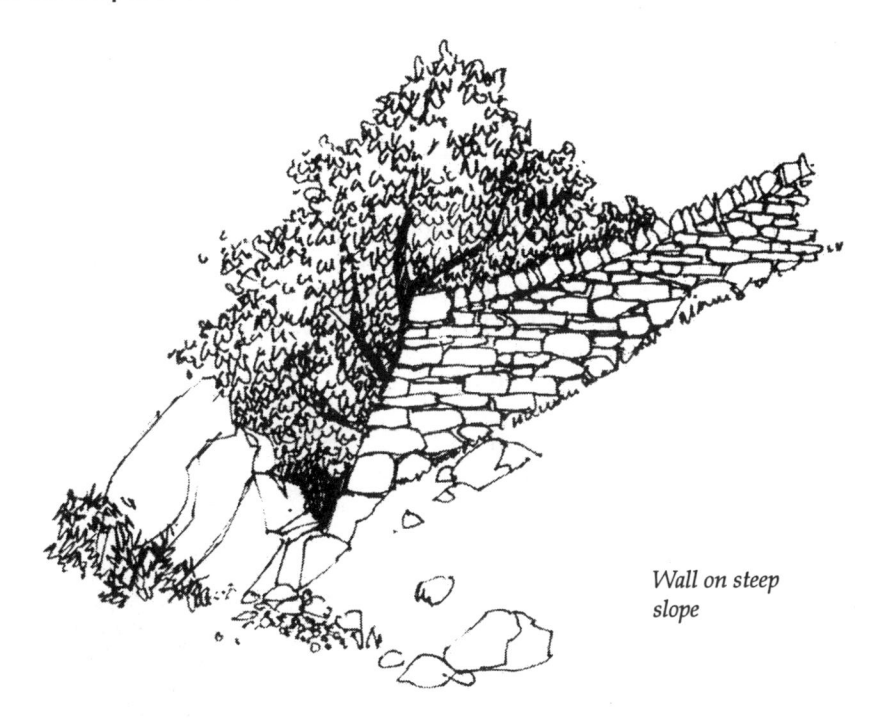

Wall on steep slope

To accommodate the slope the footings are stepped and the 'lifts' coursed horizontally to fill in to the copestones. It's a strange feeling at first building in such a way, completing say a metre of stepped footing and then immediately filling to the cope horizontally before going on to the next metre of stepping - and so on ...

The other important difference concerns the copes. At the lowest part of the wall, at its beginning, the cope should be anchored securely onto the courses below. In most cases a substantial stone will suffice but on precarious slopes the cheek end of the wall will have to be so designed as to form a secure cup to hold the end copestone. It may be that the top layer will have to resemble a long through, sunk endways into the wall from the cheek to support and tie the wall at the end.

The reason for this is that all the resulting copestones will lean against this end copestone and remain vertical whatever the angle of the slope. Absolute security at this cheek end and cope is essential. Those are the principles that apply in general to building on sloping ground but much improvisation is usually necessary to meet the varied conditions of slope and availability of stone to complete a secure wall.

4: Curve

In this book I describe the most beautifully curved wall I have seen as the one which Steven and his team built for Andy Goldsworthy in America; the 'walking wall.' It includes in its length a wide variety of curves that follow its snaking route over 360 metres.

A curved wall has exactly the same cross-section as a straight wall but the arrangement and choice of stone, is determined by the tightness or otherwise of its curve. This may be self evident, but in practice, especially with the construction of a very tight curve, this can cause innumerable problems.

The choice of stone for the inner and outer sides of an elongated curve may vary little with those of a straight wall. With a tight curve the stone has to be selected and graded as carefully as an arch (see under *Arch* in this chapter).

Curved wall

Wedge-shaped stones have their longer face placed on the outer curve of the wall and their narrower face on the inner curve. Quite often, narrow rectangles, 'fingers' of stone are used for the inner curve, their length into the centre contributing properly to the overall solidity of the wall.

On a long curve, the random stones are arranged so that their straight faces create the curve by the their tangential placement to it.

A series of continuous straight edges butted up to each other simply following the curve.

Most wallers will prepare the curve as they dig out the ground for the footings. The line of the finished curve is determined at the footing. Some wallers will simply 'eye' the curve as it is being prepared, others will measure and describe it depending largely upon its function, the landscape in which it is to be built and the waller building it. Andy Goldsworthy pegged out every metre of his route between trees and the wall had to follow this.

The biggest problem with a curve is keeping the batter consistent (especially on a tight inner side), keeping the courses parallel with each other and the ground, and maintaining a uniform line. On a tight curve the copestones will either be chosen for their wedge shape, or will be thin enough to 'fan' around it.

A beautifully constructed curved wall, following the natural contours of an open landscape, is, in my view, one of the most satisfying sights still available to us in the British hills.

5: Cheek end

These are created at the end of a wall, seal and stabilise it, and may butt up to a gatepost or other building. Often they act as a pillar at an open entrance.

In sealing the end, the waller is conscious of the cross-section of its design. Usually the stone is sorted and prepared before hand, and measured to follow accurately the batter of the wall.

Cheekend

Stones thus selected will tie well back into and across the wall, will be laid on a two-on-one, one-on-two basis, and will be covered with a substantial copestone.

A cheek is the end stop to a wall and not only has to look good, but has to be secure and solid enough not to peel away or be easily dislodged. The care necessary to achieve this is reflected in the fact that it usually takes twice as long as a metre run of wall to complete.

6: Corner

The point where two walls tie into each other. Whilst it is fairly self - evident how the outer wall develops with alternate courses going lengthways and endways over each other (always going 'into' the wall) the inner part of the corner needs special attention.

Corner

It is important to make sure that there is enough room left from the outer stones to cover the inner joints at the line of the corner. If this is observed then the corner should truly be the strongest part of the wall.

The only other note is that pins or batter frames should be set at the end of both walls to maintain a true batter in both directions where the walls meets at the corner.

7: Lunky, Hog Hole or Waterpen

This hole at the base of a wall is designed allow a passage through it for sheep, or water from a drain or small stream. It consists mainly of a hard stone base, two cheek type walls, and a 'lintel' of throughstone proportions, which spans it. Sometimes, lunkies are arched.

The footing at the base should ideally span it and be substantial. This is designed to prevent erosion of the subsoil as a result of the passage of sheep and water. The two walls either side, similar to the design of cheeks, should be well tied back into the wall, have a one-over-two, two-over-one, courses from the footings, and a 'lintel' stone which not only spans the hole but goes well into the wall on either side.

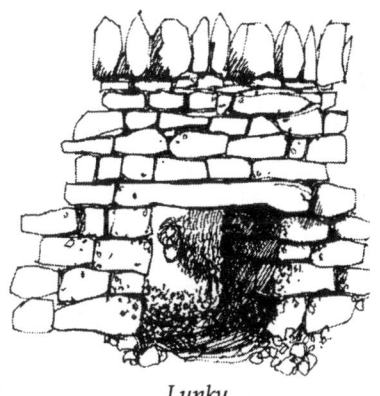

Lunky

Again this feature requires extra time and care in its preparation of the stone used in its construction.

8: Stiles

There are a number of stiles and variants that can be found in Britain. I shall describe two that are prevalent in Cumbria and the Pennines.

The simplest stile consists of two or three substantial throughs that protrude on either side with enough showing to support a large boot (about 22-25 cms). These are angled from the footing along the side of the wall and have a vertical step of around 25 cms. At the top of the wall the 'climb over point' usually has a flat, secure, substantial cope, which can cater for the climbing traffic without becoming dislodged.

Some stiles have a walk-through design at the top of the wall. These tend to be 'v' shaped and are constructed again on the same principles as a cheek end. The steps up to this step-through feature are exactly the same as described above.

9: Steps

These are not a requirement of the Master Craftsman Certificate. The 'stepping' is determined mainly at the digging-out stage of the preparation for the footings. The height of each step should be no greater than 22 cms and the width of the step no less than the length of a large boot. Sometimes the step can be a metre long depending upon the angle of the slope and the geological composition of the terrain

The foundations for the step should be substantial at the exposed end and the filling packed tightly behind it - a bit like a small retaining wall. It is important that the stepping cover stone is secure devoid of movement, and really does cover the construction beneath it to prevent it from being flushed away during inclement weather, and also to protect it from the constant traffic using it.

10: Pillar

This is usually square or round in cross-section. The principles of its construction are the same as for the corner or tight curve. The stone is laid tightly, coursed and covered in exactly the same way.

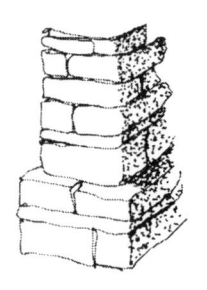

Pillar base

However, it takes special skill to keep the batter even on a conical pillar and also to keep each corner vertical on a square pillar. Most pillars will benefit from the use of a template made of wood or metal to rest on its cross-section as each course is built. Often pillars are constructed in a decorative way around major steel or concrete cores when the pillar has to support a roof.

11: Arch

The most decorative feature is the arch, which is often supported by two pillars (see above). Firstly, once the radius of the arch has been determined then a semicircular wooden frame has to be made and set in position to support the arch during its construction. Quite often old tractor tyres, being substantial and readily available enough to do the job, determine the size of the arch.

Arch – for lunky

Once the "frame" is in position and the stone carefully selected the wall is built to this frame until the uppermost or keystone is ready to fill the final gap in the arch. This keystone is then lowered or wedged into position and the rest of the wall above and either side of the arch completed in the usual way.

Pillars and arch

Usually the arch width will be determined by the width of the single stones spanning it. If it has two sides they have to be covered to prevent the filling falling through.

And finally the DSWA offers the following notes regarding the health and safety approach it expects from all wallers.

'Observance of health and Safety Regulations. Methodical approach to work. Aptitude for and attitude to the job in hand, particularly to safe handling of heavy stones and instruction to others giving assistance if required. Consistent display of skill and safe use of hammer. Keeping working area clear of obstacles. Leaving any surplus stone tidy in heaps. Appropriate working gear particularly boots and safety spectacles/goggles especially on stone that splinters. Due regard given to the preservation of wildlife and habitat within, on and beside the wall.'

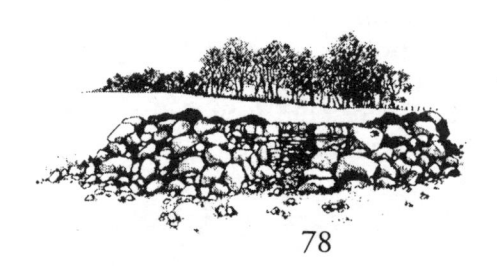

CHAPTER FIVE

Walls and walling

A survey of styles and condition in Britain

Introduction

In 1994 the Countryside Commission contracted the Agricultural Development and Advisory Service (ADAS) to survey the condition of dry stone walls in England. This was the first time that such a survey had been conducted in Britain.

The survey focused on areas that were higher than 100 metres above sea level. From this vast landscape, 700 one-kilometre National Grid Squares were randomly selected for surveying.

All the walls in each sample square were individually examined and assessed as being in one of six condition categories devised specifically for this project in conjunction with the DSWA. The categories and results are:

Group	Description Wall length found
A:	Stockproof and in excellent condition (4%: 4500kms)
B:	Sound and stockproof with minor defects (9%: 9800kms)
C:	Major signs of advancing or potential deterioration (38%: 42,700kms)
D:	Not stockproof, and in early stages of dereliction (20%: 22,800kms)
E:	Derelict (12%: 14,000kms)
F:	Remnants (17%: 18,800kms)

The survey considered and compared conditions of walls found at different altitudes; in areas of grassland, arable land and woodland; Less Favoured Areas (LFAs) of relatively poor agricultural potential; Environmentally Sensitive Areas (ESAs) of landscape which encourage governmental payments to encourage owners to preserve the traditional character of their area; and the National Parks.

I make this list because one of the fascinating results of the survey indicates that in all these areas: ' ... there is little difference in the condition of the dry stone walls except in woodland where they appear to be in poorer condition although this may be due to the fact that such walls are relics of a former land use other than woodland'.

Even more fascinating (and somewhat disturbing) is the fact that at the time of the survey, in landscapes managed by what one would regard as being 'dry stone wall sympathetic' bodies such as the National Parks, and ESAs where land management payments are available to conserve the traditional landscape character, the condition of the walls was found to be the same as those areas which have no such incentive.

The implication is that if this state of affairs is not reconciled quickly 'the landscape impact of their future decline could be very significant and that the rate of wall loss would be expected to accelerate if the condition of walls continues to worsen... and without concerted action now, more walls may be lost in the final fifteen years of the twentieth century than were lost in the previous 40 years'.

A condensed summary of the full report is available from the Countryside Commission. (See Appendix 3)

Walling Styles in Britain

There are a number of individual pamphlets on walling styles, published by the DSWA, which are readily available on request. It would serve no purpose simply to replicate these. I do however think it useful to include a general list of all the styles that prevail in Britain for the purposes of comparison: to see at- a- glance how they differ in design and structure from region to region and offer some of the reasons for this.

The accompanying notes and drawings indicate a general description of the indigenous landscape, the common stone type, logical reasons for the local style of wall construction, and the recognisable style of the copestone arrangements.

I have bracketed areas of common style together, and I consider only the traditional field wall designs. I begin with Steven's own homeland, Cumbria.

Cumbria

The geology is varied and the types of stone encourage different detail in construction. In the south of the area the stone is mainly carboniferous limestone, small and irregular in shape.

Travelling north you will encounter areas of volcanic rock (including the green slate slabs so characteristic of the northern lakes), green-granite boulders, and in the extreme north the finely coursed Skiddaw slate.

The walls however are characterised by having two sets of throughs and sloping copestones. The only main variant to this is in a tiny area around Penrith (and especially at Langwathby – see below) where the throughs are made of red sandstone, sandwiching granite boulders, and the copestones being set vertically.

Langwathby wall

Pennines

This is a huge walling area that embraces Northumberland, Yorkshire, Lancashire, Derbyshire and parts of North Staffordshire.

The two main indigenous stone types are carboniferous limestone and millstone grit. The limestone tends to be small, irregularly triangular with a general lack of throughs. The style in consequence is usually broad based and random in wall and cope.

The gritstone is more regularly cuboid, sticks together like glue, the walls are narrower, more elegant, and the vertically set copes sometime dressed in a semi-circular shape. Throughs are plentiful and there are usually two rows of them in typical field walls.

Cotswolds

The distinctive Cotswold walls are made of Jurassic limestone which is soft and, unfortunately, prone to erosion from weathering and road salt. They are however stunningly attractive walls when put together by master craftsmen.

Building stones are small, throughstones almost non-existent and copingstones in general difficult to find. These conditions don't help the waller to build a sturdy wall. In consequence walls tend to be smaller in height, and the courses hammer-trimmed to fit more closely together. Sometimes cement is added to the middle courses to tie the sides, in the absence of throughs, to support and seal the random, vertically set copestones.

Cotswold wall

Wales, Devon and Cornwall

Whilst there are numerous examples of the traditional constructions already described both areas have a characteristic feature which is best described as a clawdd or stone hedge.

It is a two-sided wall with an infill of earth replacing the familiar hearting of small stones. At the top of the wall the 'cope' is either stone or - more characteristically- turf or brushwood.

Cornish Wall

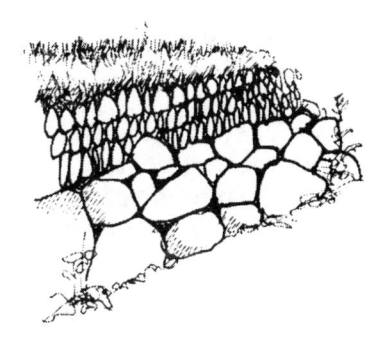

Clawdd

The base width is usually equal to the total wall height, the cross section having a concave design to compensate for the settling of the earth infill with time and its tendency to push the lower courses outwards.

The courses of stone are laid vertically and are usually coursed. The distinctive herring-bone patterns seen in the upper courses, particularly in Cornish hedges (left), are so designed to provide a better rooting structure for the vegetation above them. The other main feature of these walls is that the stones tilt down to the centre to encourage rain to drain to the roots of the vegetation buried within rather than off the wall.

Scotland

The Galloway dry stone walls or dykes display the most distinctive style of walling to be found in Scotland. The stone is dense, heavy and often rounded boulders of granite. The lower courses are laid as level as possible despite the variety of size and shape in a familiar double wall with filling style, tied with throughs at the halfway point, and a coverband beneath the cope.

Often, the lower wall, up to the through is only about 0.9 metres high is then tied with a coverband, and two layers of large, single, rounded stones spanning the wall form a kind of double cope. This style often combines a double (at the bottom) and single wall at the top.

Galloway Dyke

A boulder dyke is made from single stones piled on top of each other with the largest stones at the foot of the wall. The stones are set across and into the gaps of the stones beneath, and their reduction in size as the wall lifts must maintain the 'A' shape of the batter.

Ireland

The Irish walls follow the styles of their other British counterparts, but occasionally various distinctive patterns of construction emerge, again because of the type of stone available in a particular region. The illustration below describes what looks like an almost random, 'pack-of-cards' vertical style, where the waller has 'walked' the huge slabs and leaned them against each other at the base and placed the more liftable stones on top to a convenient height, in a single-width structure which is both strong and practical.

Irish wall

Others

There are inevitably a number of modifications to all the above styles that can be found up and down the landscape of rural and industrial Britain. These reflect the ingenuity of local craftsmen and their attempt to utilise the particular geological conditions that prevail.

However, there is one style which, though not constructed in the stone-on-stone method, does represent another way of marking and enclosing the borders of the land.

Flag Walls

These are constructed with flat slabs (flags) of stone, some are trimmed to butt next to each other vertically, others overlap each other and sometimes metal plates are used to secure neighbouring slabs: all are set into the earth sufficiently deeply to be self-supporting – a bit like a simple gravestone. Examples of these styles can be found in Caithness in Scotland, Hawkshead in the Lake District, in Lancashire and occasionally in Cheshire.

Cumbria flag wall

Welsh slate fence

Occasionally, as in the example from North Wales, the flags are set wider apart and tied with a double length of wire twisted close to the top around each vertical slab of slate.

CHAPTER SIX

In Conclusion

It is interesting to listen to the comments of other wallers regarding Steven's work. I have yet to meet a fellow professional who does not regard his skill as being very special. Bill Noble, an equally fine waller, who has shared sites with Steven on a number of occasions both in work and competition, regards him as being the finest waller of his day. He remembers particularly the first walling project in America (described earlier in this book) and the particular problems of the local stone which was clumsy and tricky to work, and how Steven managed somehow with the same stone, to make his stints of wall look 'tidier and tighter' than the others around him.

It was also in the same company that Steven heard the now immortal phrase from a local waller that 'a stone is a stone'. No matter what its shape or structure, providing it holds together well over decades of weathering, it is a suitable substance to construct with. Steven manages to build solid walls with any stone and make it look good.

Often wallers complain about the condition of the stone with which they are expected to build. As Grand Prix competitors travel the regions they can often be heard criticising the local stone defending the stone they are used to working with on their own patch. To voice such complaints to Steven is to receive a shrug of the shoulders and the standard reply 'a stone is a stone'.

I have often asked Steven what it was that made dry stone walling so special and still challenging to him. He has no pat answer. All he can say is that stone and walls are for him a way of life that has grown out of his experience of living and working. He can't remember a time when he hasn't looked at walls without instinctively assessing their general context, construction and design, wondering about the stone, who had built it, how would he have done it differently.

Whilst he feels nothing romantic about the physical challenge and demands of walling he doesn't mind the solitary aspect of the job even though he wouldn't like to be on his own every day, especially given the inherent problem of working alone - from a health and safety point of view.

He likes travelling to different places and meeting different people and walling challenges. It's the variety at this moment that is particularly attractive. Although, all the time he accepts a new and particularly exciting challenge he never turns away the basic ' bread-and-butter' repair work of gaps in field walls, especially for those local farmers who supported him in his early days of self-employment.

The best day's walling I shared with Steven was such a session, gapping on the Cumbrian fells in the middle of winter. I had driven from Leeds, for a meet at Tebay whilst it was still dark, and continued the journey to the site, eventually joining Steven and following him in convoy as the dawn broke.

The site was situated at the top of a hillside pasture, and we were able to drive up to it confidently only because of the rock-frosted grass. As the sun rose, the sheep revealed themselves and inquisitively followed their long shadows to the gap we were preparing to repair. Our presence must have suggested a familiar ritual of feeding, but once they realised we were not going to spread bales of hay, they shuffled away.

The white, crystalline slopes glistened, and the heart of the sun touched the refracted outline of the frozen tumble. There was a stillness that softened the familiar clunk of our movement of stone.

That was it. Simple, physical and quite breathtakingly beautiful. The man, the wall, the landscape, the day: in harmony.

APPENDICES

1: **Glossary of terms**
This list includes a separate list of those terms already defined in footnotes in this book, giving the contextual page reference.

2: **Bibliography**
The bibliography lists books both in and out of print that relate to the craft.

The Dry Stone Walling Association publishes a number of leaflets on specific aspects of the craft, such a technical specification and region styles. Details of these and books currently in print and available by mail order are available from the Dry Stone Walling Association. Please enclose a stamped, self-addressed envelope with your request for details.

3: **List of Organisations**
This list was current at the time of going to print and shows organisations and agencies that have an interest in the rural situation generally or dry stone walling in particular.

4: **DSWA Statement of Policy**

5: **DSWA Craftsman Certification Scheme**
A brief outline of the scheme and tests available. The scheme undergoes a periodic every two or three years to meet the needs of wallers and dykers.

APPENDIX 1

GLOSSARY OF DRY STONE WALLING TERMS

"A" FRAME: A wooden or metal frame used as a guide when building. Also called a batter frame.

BATTER: This is the inward taper of the wall from base to top, the vertical angle of the sides between footing and cope. See footnote 12.

BEE BOLE: Niche constructed in a wall to house and provide shelter for traditional bee skeps.

BROCH: Scottish fortified, circular construction in dry stone thought to have originally been built as a defence, although recent research suggests these were homes.

BUCK & DOE: Top or cope stones alternating with large and small stones. Also called castellated, cock & hen, etc.

CAMBER: the even tapering of the wall along its length.

CONSUMPTION DYKE: Wall built with stone to clear the land and which is especially wide. Also called "clearance wall" and "accretion wall".

COPE STONES: (Top stones) The stones along the top which tie, give weight and protection at the top of the wall. Also called "cams", "tops", "toppers". See footnote 9.

COURSE: Horizontal layer of stones placed in a wall.

COURSED RUBBLE WALL: Rubble wall laid in courses.

COURSING: The method of keeping each layer of similar depth stone to the same level. See footnote 21.

COVERBAND: Large flat stones placed across the width at the top of the wall (in some regions) to form the base for the copestones. See footnote 24.

CRIPPLE HOLE: An opening in the lower half of a wall, usually rectangular, to permit sheep to pass through but to retain cattle. Also called lunky, lunkie, lonky, hogg hole, smoot hole, thawl or thirl. See also "Waterpen". See footnote 28.

DOUBLING OR DOUBLE DYKING: A term used for a dry stone wall built with two faces of stones, packed with hearting between.

FACE: The exposed side of a wall.

FACE STONE: A stone that forms part of the outer face of a wall or dyke.

FIRST LIFT: The building of the wall from foundations to first row of throughstones. See footnote 22.

FLAG: Technically sandstone that is thinly bedded and readily split into slabs. Often used to describe any flat slab of stone.

FOUNDATION: the first layer of large stones in the base of the wall, also called "footing" or "found". See footnote 6.

FREESTONE: term used to describe stone with no clear cleavage lines but will break in any direction.

GALLOWAY DYKE: A wall or dyke with its lower third "doubled", and upper two thirds in single walling.

GAP OR GAPPING: A breach in a dry stone wall. Gapping is the repair of the same and the "gapper" is the waller or dyker who carries out the repair. See footnote 5.

GRAND PRIX: A grouping of local competitions to make a national circuit. See footnote 1.

GRIT, MILLSTONE GRIT, GRITSTONE: Dense, hard sandstone.

HEARTING: The small stones used as filling or packing in a double wall. See footnote 8.

HOGG HOLE: See Cripple Hole.

INSIDE OUT AND OUTSIDE IN: Building with the length of stone into the wall. See footnote 26.

JOINT: The line between adjacent stones within one course of a wall.

LIFT: A lift is the construction of the wall or dyke between the foundation and throughs; throughs and copes. See First Lift, Second Lift.

LINES: Cotton or similar line attached to pegs or frame used to maintain level of each course. See footnote 19.

LUNKY/LUNKIE: See Cripple Hole.

MARCH DYKE: Scottish dyke forming major boundaries between estates.

MONOLITH: Term often used for a single large stone set in the ground. Occasionally used to describe a stoop. See footnote 17.

OOLITIC LIMESTONE: The soft limestone found mainly in the Cotswold region of Britain.

PINNINGS: Small, usually tapering stones used to wedge building stones firmly in place.

PINS: Steel rods driven into the ground at the footings of the wall to mark the vertical cross-section from footing to cope. Can be used instead of batter frames, particularly along the length of a wall. See footnote 20.

RANDOM: Wall built without obvious lines of coursing using irregular shaped stone. See footnote 11.

RANDOM RUBBLE WALL: Rubble wall laid uncoursed.

RETAINING WALL: A dry stone wall built into the cut face of a bank to prevent the soil from moving down the slope.

RUBBLE WALL: A wall built with stones of irregular shape & size

SCARCEMENT: Usually found in Scottish dykes, the footings are set wider than the first course of the wall.

SECOND LIFT: Building from first throughstones to coverband, or second throughs.
See footnote 23.

SINGLE DYKE: A wall built with single stones going the width of the wall. Also called a boulder dyke.

SMOOT: Usually describes small passages constructed in the base of a wall to allow the passage of rabbits and hares, or even slightly larger for badgers. Sheep smoots (see above).
Water smoot: see waterpen.

STILE: A construction within a wall to allow easy passage of walkers through or over the wall (e.g. squeeze stile, step stile, etc.)
See footnote 16.

STINT: A measured length of wall or dyke such as in a competition. See footnote 2.

STOOP/STOUP: Upright stone set into ground, often at a wallhead to form a gatepost.

STRIPPING OUT: The careful pulling down of a wall, stone by stone, and laying out of the varying construction stones in readiness for rebuilding.
See footnote 25.

THROUGHSTONES: Heavy, large stones placed at regular intervals along the wall to tie the two sides together. Often simply called "throughs".
See footnote 10.

TIE: A stone set above two, often through the width of a wall, to tie two sides together (e.g. throughs and coverstones)
See footnote 7.

TRACE WALLING: Incorrect placing of stones with their length along face of wall rather than into the wall for strength.
See footnote 27.

TOPSTONE: Cope stone.
See footnote 13.

TUMBLE: A short length within a wall that has fallen or tumbled.
See footnote 4.

WALLHEAD: Vertical end to a length of wall. Also called "cheekend". See footnote 3.

WATERPEN: A construction to allow passage of water beneath a wall often similar in construction to a lunkie or smoot.

WHINSTONE: Term used for variety of hard rock such as greenstone, basalt, chert and quartsoze sandstone.

A definitive glossary of terms is in the process of being collated and cross-referenced by the DSWA and should be available for purchase in leaflet form sometime in the early of the Millennium Year.

APPENDIX 2 - BIBLIOGRAPHY

BUILDING & REPAIRING DRY STONE WALLS Author R Tufnell. Published by DSWA. ISBN 0-9512306 2 X. *Popular publication giving basic information on construction of dry stone walls (double-style). Written as a beginner's guide, straightforward line illustrations clearly show the right way to lay stone. First published by DSWA in 1982, extensively revised 1991.*

BUILDING SPECIAL FEATURES IN DRY STONE Author R Tufnell. Published by DSWA. ISBN 0 9512306 1 1*First published in 1990 this booklet has proved most popular as a source of basic information on constructing many features from stiles and lunkies, to raised flower beds, retaining walls & shooting butts. Reset and reprinted 1997.*

BETTER DRY STONE WALLING Author R Tufnell. Published in 1991 by DSWA. ISBN 0-9512306 3 8. *An insight into the techniques required for that all-important professional finish. Includes double and single style walling, work on sloping ground, etc. Clear line drawings throughout.*

CREATING A NATURAL STONE GARDEN Authors: R Tufnell, I K Dewar, P Webley, M Ribchester. Edited for DSWA. Published 1996 by DSWA. ISBN 0-9512306 4 6. *This is a book of ideas for using dry stone in the garden and landscape. Excellent drawings showing ideas for planting and use of stone within the garden.*

DRY STONE WALLING Author: Col F Rainsford Hannay. Published by Stewartry DDC. ISBN 0-9502623 0 7. *First published in 1957, written when there was a definite waning of interest in the craft. It has stood the test of time and provides a valuable insight into the history and construction of dry stone walls around the country. Reprinted 1976, 1999.*

DRY STONE WALLING Author Alan Brooks, revised by Elizabeth Agate 1996. Published by BTCV. ISBN 0-9501643 5 6. *A practical handbook containing information about history, construction, conservation and legal aspects of dry stone walls.*

MAKING STONEWORK Authors: Marie Hartley & Joan Ingilby. Published by Smith Settle. ISBN 1 85825 080 3. *Photographs and notes of craftsmen and their work. Mostly stonemasons, interesting pictures "in the field". Good book for the collection.*

DRY STONE WALLS Author: Lawrence Garner. Published by Shire. ISBN 0 85263 6660. *A short guide to dry stone walls providing basic history and brief construction information.*

WHAT'S ON A WALL Published by SCE Ltd *A valuable ecology guide with useful recording sheets for use in the field.*

STONE Author: Andy Goldsworthy. Published by Viking 1994. ISBN: 0 670 85478 6. *"Harnessing the image in the book to the concept of stone has encouraged Goldsworthy to question his (and our) perceptions of time, stability, change and impermanence".*

WOOD Author Andy Goldsworthy. Published by Viking 1996. ISBN 0 67 087137 0. *Andy Goldsworth both parallels and extends the themes in the previous book STONE.*

SHEEPFOLDS Author Andy Goldworthy. Published by Michael Hue-Williams Fine Art 1996. ISBN: 0 900829 00 3. *A graphic descriptions of an environment commission to build one hundred sheepfolds as placements for permanent sculptures in dry stone designed by Goldsworthy on original sites along the drover's routes of Cumbria.*

Out of Print Titles

PENNINE WALLS Author Arthur Raistrick. Published by Dalesman Books 1966. *Illustrated booklet on the walling history, craftsmen and techniques of the region.*

LAKELAND WALLS
Author William Rollinson. Published by Dalesman Books 1972. *Companion booklet to Pennine Walls with emphasis on geology and history.*

LAKE DISTRICT STONE WALLS
Author Janet Bodman. Published by Dalesman Books 1984. *Booklet on the regional styles, mainly house building.*

DRYSTONE DYKING
Author Robert Cairns. Published by Biggar Museum Trust 1986. *Reminiscences of the Lothian and Tweeddale regions of Scotland and estate work including drystane dyking.*

DYSTANE DYKING IN DEESIDE
Author Robin Callander. Published 1986. *History and techniques of the Deeside region.*

THE DRY STONE WALL HANDBOOK
Author Edward Hart. Published by Thorsons Publishers Ltd, 1980. *Book of dry stone walling techniques.*

THE STONE WALLS OF IRELAND
Authors A MacSweeney and A Conniff. Published by Thomas and Hudson, 1986. *Walls and their place in the landscape. Most of dry stone although some mortared walls.*

WEST COUNTRY DRY STONE WALLS
Author Janet Bodman. Published by Redcliffe Press, Bristol in 1979. ISBN: 0 905459 1 *Mostly concerned with mortared walls.*

DRY STONE WALLS OF THE YORKSHIRE DALES
Author: W R Mitchell. Published by Castleberg Books, Giggleswick. ISBN: 1 871064 8. *Includes sections about Enclosure Awards, History, Natural History and Farming.*

YORKSHIRE DALES STONE WALLER
Authors: G Lund, R Muir, M Colbeck. Published by Dalesman Books, 1992. ISBN: 1 85568 04. *Colour photographs with text of Yorkshire Dales scenery with walls, stiles and stone.*

WALL TO WALL HISTORY – THE STORY OF ROYSTONE GRANGE
Author: Richard Hodges. Published by Gerald Duckworth & Co Ltd, 1991. ISBN: 0 7156 234. *Detailed archaeological account of dry stone walls on an estate in the White Peak of Derbyshire.*

FIELDS IN THE ENGLISH LANDSCAPE
Author Christopher Taylor. Published by Dent, 1975. ISBN 0 460 0223. *Includes a useful history of field boundaries, including walls.*

APPENDIX 3

ORGANISATIONS

The Dry Stone Walling Association of Great Britain

A registered charity, founded in 1968, which welcomes members from all walks who are interested in, or who work within the craft of, dry stone walling. The work of the Association encompasses all aspects of the craft. A number of books and leaflets are published, including a register of certificated professional members. The DSWA operates the only tiered, national skill certification scheme devoted to dry stone walling.

The Association has active branches in most of the "walling" regions of England, Wales and Scotland. Most branches organise practice meets and training courses open to the public. On going to print, the branches are: Central Scotland, Cheshire, Cotswold, Cumbria, Derbyshire, Eden Valley, Isle of Skye, Lancashire, Mid-Lancs, Northumberland, North Wales, Pennine, South East Scotland, South Wales, South West Scotland, South Yorkshire, West of Scotland, West Yorkshire and Yorkshire Dales. A list of current honorary branch secretaries and an annual list of practice meets and courses is available from the Association.

The Association can be contacted by post through its registered address:

**Dry Stone Walling Association,
c/o YFC Centre,
National Agricultural Centre,
Stoneleigh Park,
Warwickshire CV8 2LG**

Other Organisations

The Friends of the Lake District
Murley Moss,
Oxenholme Road,
Kendal,
Cumbria LA9 7SS

Countryside Agency (England)
John Dower House,
Crescent Place,
Cheltenham,
Gloucestershire GL50 3RA

Scottish Natural Heritage
Battleby,
Redgorton,
Perth PH1 3EW

Cyngor Cefn Gwlad Cymru
Plas Penrhos,
Ffordd Penrhos,
Bangor,
Gwynedd LL57 2LQ

Crofters Commission
4-6 Castle Wynd,
Inverness,
Scotland IV2 3EQ

Council for the Protection of Rural England (CPRE)
Warwick House,
25 Buckingham Palace Road,
London SW1W 0PP

APPENDIX 4

DSWA STATEMENT OF POLICY

The objects of the Dry Stone Walling Association of Great Britain are to preserve, improve and advance knowledge of the craft of dry stone walling for the benefit of the public. The Association is a registered charity and a voluntary organisation with local branches in many parts of Britain.

- The Association recognises that the craft is carried out by professional wallers/dykers, working farmers, farm workers, amateurs and conservation volunteers. All need encouragement to improve the quality and quantity of their craft.

- The Association works to resist the unnecessary destruction - either by design or neglect - of existing dry stone walls.

- The Association encourages the maintenance of existing walls, the rebuilding of damaged or neglected walls and the building of new walls by promoting the craft to landowners, tenants and land agents; to the authorities responsible for National Parks, water and highways; and to architects, planners plus other private and public bodies.

- The Association encourages the Government, the European Community and Government bodies to provide adequate grants for the construction, maintenance and management of dry stone walls. For these to be effective (and to encourage the uptake of these grants) the schemes should provide sufficient recompense for the work done.

- The Association works towards the provision of adequate training schemes at reasonable cost for all participants in the craft. The Association operates training programmes in the craft through its local branches.

- The Association operates a practical, craft skills certification scheme to ensure adequate standards and encourages all individuals and organisations to recognise and use the *DSWA Craftsman Certification Scheme*.

APPENDIX 5

DSWA CRAFTSMAN CERTIFICATION SCHEME

The **Dry Stone Walling Association of Great Britain** encompasses all aspects of the craft in Britain today. Part of the Association's work involves operating a national series of progressive, practical tests leading to the *Master Craftsman Certificate* in dry stone walling. Those wishing to participate should read the full booklet *DSWA Craftsman Certification Scheme --Introduction and Schedules*, which is available from DSWA.

The scheme, established in the early 1980s, provides recognized skills certification for wallers and dykers giving guidance to employers on the ability of the individual. The *Craftsman Certification Scheme* provides a range of tests to meet the needs of today's working waller and dyker in the wide range of work situations: from straight-forward repair work to extensive landscaping projects involving many features.

The *Craftsman Certification Scheme* is designed to ensure candidates achieve the highest standards in the craft. Tests are open to all individuals, whether or not a member of the Association and are in four levels:
- **Initial** covers the basics of the craft – how to repair a gap;
- **Intermediate** includes the building of a wall with a cheekend or wall head;
- **Advanced** involves retaining walls;
- **Master Craftsman** covers the building of various structures to a high degree of finish.

In addition, there is a category for *Regional Styles*. Currently there is a test for the Galloway style, open to those of Intermediate Certificate standard. It is anticipated this section of the scheme will encompass further regional styles as and when there is demand for recognized certification.

The Association has a group of Master Craftsman Certificate holders who have been trained in skills assessment and who regularly come together for review of standards. These are the DSWA examiners who are present throughout the practical tests and are the only people who can undertake skills assessment for the Scheme.

Local branches of the Association may operate "test days"; candidates can also arrange for tests to be carried out at their regular place of work, or as part of a training scheme. Most DSWA branches operate training weekends and practice meets for the beginner or improver. A list of branch contacts is available from DSWA.

The full booklet *Craftsman Certification Scheme – Introduction and Schedules* contains details of the requirements for each test, some of the key points used by examiners when undertaking tests and notes to guide possible applicants. Candidates, or those thinking of taking tests, are strongly advised to read the full booklet prior to taking part.